JN049102

光のトンネル
Tunnel of Light

馬岩松
Ma Yansong

現代企画室
GENDAIKIKAKUSHITSU

序

マ・ヤンソン
MAD アーキテクツ設立者、共同主宰

とらえがたさ

初めてこの場所を訪れた際、とても、とても深い、山の中のトンネルで、突き当りは空っぽであった。なぜ入ったのか分からない。出るため？　それとも、出られないと知るため？　いずれにせよ、中は何もなく、空っぽだった。外部は元々のままだが、異なる角度から眺めたことで、自分自身が変化した。

入口

すべての洞窟の入口はそこから離れるためにあり、だからとても明るく、彼方への誘惑と想像で満ちている。最も慣れ親しんだ安全な場所から離れることを決意するには、どれほどの勇気がいるだろうか。私にできることは、離れる決断を迷っている人にしばらく付き合うことである。

反射

私は鏡を気の毒だと思う。鏡にできるのは複製することだけだが、現実は、複製する必要はない。私たちは本当に自分自身をはっきりと見たいのだろうか？本当にはっきりと見ることができるのか？　夢の中は何もはっきりとは見えないが、最も深い印象が残る。見るものが真実と限らないのであれば、現実を曖昧化し、天と地の境界線を滲ませよう。

水

山間には清津峡渓谷の雪解け水が流れ、四季に彩られさまざまな姿をみせる。だが、願わくば足元に静かにとどまって、その中に足を浸し、骨を刺すような冷たさにより、日常で麻痺した私の感覚を取り戻させてほしい。

Preface

Ma Yansong, founding partner and principal architect, MAD Architects

Empty Spirit

It's my first time coming to this place, so, so deep. At the end of this pipeline through the mountains… nothing. I don't know why I came here. Was it to exit the other side? Or was it to know there is no exit? In any case, there's nothing inside, just emptiness. Outside is still the same outside it was before, but I've switched perspectives. The only thing that has changed is myself.

Cave Mouth

Cave mouths are for leaving. Their brightness, filled with imagination and temptations of distance, grant us the resolve to bid farewell to the safest and most familiar of places. What courage it takes. All I can offer is this brief companionship in the hesitation before departure.

Reflections

Such sad things, mirrors. They can only make copies, but what need have we to copy reality? Do we really want to see ourselves so clearly? Are we capable of seeing clearly? In dreams, where nothing is seen clearly, we are moved to our depths. If what the eyes see is not real, let reality become a blur. May all between sky and earth blur.

Water

Snowmelt flows year-round between the mountains of Kiyotsu Gorge, passing a thousand shifting palettes and forms, but I think it should stop here at my feet. Be still. I stick my foot in and let the piercing cold remind me of the numbness of the everyday.

無

誰もいない方がいい、自分自身でさえも。

人

ひとりが天と地の境界線に近づくと、わずか20メートル離れただけなのに、まるで別世界に行ったようである。私もその世界に入りたい。20メートル離れたもうひとつの世界の彼らを眺めたい。

自然

清津峡に向かう途中、緑、山、川、空、そして不思議な光を見た。とても美しく、他に何も加えられないと思った。できるのは自分の心の感動を描くことだけ。自然は客観的であり、感情は主観的なのだ。

Nothing

Best that there be no one, not even myself.

Someone

Someone approaches the line separating sky and earth. They're only 20 meters away, but it's like they've entered another world. I would like to enter that world. I would also like to admire them in that other world, separated by these 20 meters.

Nature

On the road to Kiyotsu Gorge, I have seen all there is to see of foliage, mountains, rivers, skies, and the marvelous light —so beautiful I feel there is nothing else to add. What I can do is relate what moves inside me. Nature is objective; feeling subjective.

清津峡渓谷を再生した「光のトンネル」

The Tunnel of Light and the Rejuvenation of Kiyotsu Gorge

北川フラム

アートディレクター、アートフロントギャラリー主宰
1946年、新潟県生まれ。2000年より「大地の芸術祭 越後妻有アートトリエンナーレ」総合ディレクター、2010年より「瀬戸内国際芸術祭」総合ディレクターを務めるなど、多数のアートプロジェクトの企画、ディレクションに携わる。フランス芸術文化勲章シュヴァリエ、ポーランド文化勲章、文化功労者などを受章。

清津峡は、新潟県の山奥にある世界有数の豪雪地である越後妻有にあり、日本三大峡谷のひとつと言われてきた名所である。そこに2018年、馬岩松の《Tunnel of Light》が誕生した。トンネル誕生の経緯は、1862年、清津峡の入り口に温泉場が作られたことに始まる。

その後、1941年に清津峡は国の名勝に指定され、1944年には上信越国立公園の一部となる。1984年に大規模な雪崩で温泉街が被害を受け、1988年に渓谷内の落石で死亡事故が発生し遊歩道が閉鎖。そして1996年、地元の住民と観光業者の要望によって建設された清津峡渓谷トンネルが開業した。

開業当初は大きな賑わいを見せたものの、次第に来訪者数は減少した。2000年から大地の芸術祭がはじまり、3年に一度のその会期中には一定の来場者があるが、あとの期間は苦戦した。特に2004年の新潟県中越地震後は落ち込みが目立った。これをうけて十日町市は渓谷トンネルのリニューアルをすることになった。その設計を馬岩松のチームMADアーキテクツに依頼したのである。

清津川に沿った渓谷の懸崖の岩の中を走っている幅約3から10メートル、高さ3から7メートル、全長750メートルのトンネルは、途中3か所に峡谷を望む開口部があり、終点は奥の峡谷全体を見渡せるテラスとなる。
MADの当初の考えは、峡谷を望む見晴所3つそれぞれを、トイレ、アート作品、レストランとし、終点を

Born in Niigata Prefecture in 1946. He has served as the general director of the Echigo-Tsumari Art Triennale since 2000, the Setouchi Triennale since 2010, and other several regional art festivals which have brought him many awards, including the Ordre des Arts et des Lettres from the French Republic, the Order of Culture from the Republic of Poland, Person of Cultural Merit of Japan and so on.

Located in Echigo-Tsumari, a mountainous region in Niigata Prefecture and one of the snowiest places on Earth, Kiyotsu Gorge is known internationally as one of the three great gorges of Japan. [Editor's note: the other two are Korube Gorge in Toyama Prefecture and the Osugidani Valley in Mie Prefecture.] It is also the place where, in 2018, a nearly derelict tourist tunnel was reborn as architect Ma Yansong's "Tunnel of Light."

In 1941, Kiyotsu Gorge was named a National Scenic and Historic District. In 1944, the gorge became part of Joshin'estu-kogen National Park. In 1984, a major avalanche struck the hot-spring resort, killing five people. In 1988, after a deadly rockfall incident, hiking trails in the area were closed. In 1996, in response to requests from residents and the local tourism industry, the Kiyotsu Gorge Tunnel was built and opened to tourists.

The tunnel enjoyed a period of popularity after opening, but tourism numbers soon tapered off. In 2000, the first Echigo-Tsumari Art Triennial was held. Visitors to the tunnel increased for the duration of the festival, but afterwards once again declined, reaching an all-time low after the 2004 Chuetsu earthquake. Finally, the Tokamachi City government decided the tunnel required renovation, entrusting the design work to MAD Architects, led by Ma Yansong.

足湯にするというものだった。しかし予算の関係で、最終的にはアート作品、レストランは断念することになった。

MADはトンネル内部と新たに設置したエントランス施設のいくつかのポイントに、自然の5大要素（木、火、土、金属、水）をモチーフとして展開し、人間と自然をつなげようとした。トンネルの終点、パノラマステーションは圧巻だ。トンネルの内壁を覆う半鏡面仕上げのステンレススチールが、渓谷の風景を映しこみ、それが浅いプールに張られた水にも投射され、自然の幻影が無限のパノラマとなって広がるのである。

建設にあたっては、新潟県内の設計事務所が元請けとなり、全体のコーディネートを私が代表をつとめるアートフロントギャラリーが受け持った。この間の感想で言えば、MADのパートナーのひとりである早野洋介が複雑な条件のプロジェクトをよくまとめ上げたということに尽きる。出来上がりは写真で見ていただくしかないが、それは訪れたものだけが味わえるサイトスペシフィックなものだった。その体験は大きな評判を呼び、2018年の大地の芸術祭期間中は、連日大渋滞がおこるほどの人気となった。数字で言えば、50日間の会期中8万人、会期翌年の2019年も1年間で30万人が訪れている。

しかし、大切なのはこのトンネルが与えた日本の美術・文化・観光への決定的な影響である。日本の文化庁は、東京オリンピック・パラリンピック開催を契機に、日本の文化振興を目指し、「日本の美」を体現する文化プロジェクトを体系的に展開する「日本博」を2019年に創設し、「日本人と自然」を総合テーマとして掲げた。そのメインヴィジュアルのひとつとして清津峡トンネルの映像が使用されたのである。作家は中国人であるが、日本の文化の「自然との親和性」を象徴するものとして選ばれたのだ。山水都市を理念として設計する馬岩松が、自然を媒介として大地の芸術祭、日本の文化的特徴とつながったのだ。思えば、私たちは、その自然とのつながりを中国の文化から1500年の長きにわたって享けてきたのだった。

最後に、MADの建築について考えていることを記し

750 meters long, and varying three to ten meters in width, and three to seven meters in interior height, the tunnel passes through Kiyotsu Gorge, linking three viewing platforms where tourists can look out on the surrounding scenery. The end of the tunnel offers a vantage point overlooking the entire gorge.

The original plan by MAD Architects was to convert the viewing platforms into a restroom, an art space, and a restaurant, while the end of the tunnel would house a foot-bathing pool. However, due to budget limitations, plans for the art space and restaurant were abandoned.

Ancient Chinese philosophers understood the formation and interactions of the natural world in terms of the five elements—metal, wood, water, fire, and earth. MAD Architects took inspiration from this theory to re-envision the facilities at the tunnel entrance and at key points within the tunnel, looking for ways to bring visitors into closer contact with nature. The master stroke of the design is the viewing platform at the end of the tunnel. Polished stainless steel semi-circles cover the walls, reflecting the exterior scenery of the gorge into the tunnel and across the surface of a shallow pool in the floor, creating a seemingly boundless magic mirror through which one can observe the natural scenery outside.

Niigata Prefecture's Green Sigma Design Studio acted as contractor to implement all aspects of the renovation, and I, through my role at Art Front Gallery, acted as their representative and project coordinator. If you were to ask what impressions of this project I have formed over the past few years, I would have to give special mention to Mr. Yosuke Hayano, a principle partner at MAD Architects, who overcame highly complex conditions and countless obstacles to bring this design to fruition. Although many of you will only view it through photographs, the beauty of this work of art is dependent on the special character of open spaces, and must be seen in person to be fully appreciated.

たい。

MADの建築では、外界と建築の内部空間を限りなく近づけているようにしている。そこには皮膜一枚で押し返しあう融通無碍の関係があるだけである。

「虚実皮膜」という言葉が日本語にはある。きっと中国語にもあるだろう。裏になるか表になるかは全くの偶然であり、表裏は裏腹である。虚実は常に反転し、何が真実であるかはわからない、という謂いである。それは、世界のすべては自然に内包されているということだ。

距離をとってMADの建築群や計画模型を眺めれば、それはあたかも中国古来の山水画であるかに感じられる。あの堅牢な大地を水が流れ、風が通り、削り取られた山々は、水や空気の重層や動きによって霧や靄となり、形あるものの輪郭を曖昧にしている。それは個別な山容、仏閣、家、人の動きを含めて、万象はすべて自然のなかにあって、自然をかたちづくるものという自然観のあらわれで、MADの建築と同じように思える。

近代建築はひとつの空間を、人の動き、光と風の動きなどの検討をして、部分から全体を構築した。さらに世界中の集落個々の成立の調査により、場所によって大切にするものにかなりの違いがあることがわかるが、これも建築上の要素としてデータに入れられていった。ただ一般的に言えば、社会的な制度に制約されていて新しい展開はできにくいようだ。

さて、ホテル、高層の住居を抜かせば、私たちが写真集で見るMADの仕事のほとんどはパブリックなものだ。馬岩松はこれらの設計を通して現代の価値観に挑戦しているように見える。それが皮膜を使った表裏反転の世界である。個体が外界と接することにより外部が内部に、内部が外部に変わる、化ける空間。あるいは自然そのもの。それは鉄骨とガラスのカーテンウォールが重力に逆らう限り高層化せざるをえないという宿命から自由になりえないことを自覚した近代建築から、離陸する方法なのではないか。

思えばMADの初期の作品「フィッシュ・タンク」の

Upon completion, "Tunnel of Light" was showered with critical praise. The subsequent surge of public interest caused major traffic jams for the duration of the Echigo-Tsumari festival. Over the course of the 50 day festival, statistics show the tunnel received 80,000 visitors. Over the following year (2019), numbers swelled to over 300,000.

More importantly, however, the tunnel has had a major impact on Japanese art, culture, and tourism in general. In an effort to leverage the upcoming Tokyo Olympics and Paralympics as a means to revitalize Japanese culture, Japan's Agency for Cultural Affairs has begun promoting the Japan Expo project. Under the general theme of "Japan and Nature" the agency has organized campaigns showcasing Japanese aesthetics with "Tunnel of Light" serving as a primary showpiece.

Thus, the work of a Chinese architect has become a symbol of Japanese culture's intimate connections with the natural world. Indeed, there is a thread linking Ma Yansong's "Shanshui City" design concept, the "nature as medium" concept of the Echigo-Tsumari Art Triennial, and the unique character of Japanese culture. Upon reflection, we find that Japan's understanding of the relationship between humanity and the natural world is a Chinese cultural import that we have been enjoying for the past 1,500 years.

In the space remaining, I would like to discuss my view of the ideas inherent in the work of MAD Architects. In a MAD design, interior and exterior spaces are tightly conjoined, as if by a thin membrane that enables uninhibited exchange.

The "margin between the real and unreal" is a Japanese aesthetic concept, but perhaps something similar exists in China, where distinctions between front and back, inner and outer, are considered arbitrary. Inner is outer, and outer is inner. Even reality and fiction can trade places, making it difficult to accurately define what is real. In other words, this principle extends to all phenomena.

なかで金魚は皮膜のなかで重力を気にしないで泳いでいる。今も彼らの事務所の中に卓球台と並んで大切に場を占めている金魚鉢はMADの出発点を示していて微笑ましい。

If you view one of Ma Yansong's buildings or architectural models from afar, you feel you are admiring an ancient Chinese ink landscape. Streams flow over stones, wind blows over jagged peaks, water collides with air to form mist. Outlines blur. Valleys, shrines, and huts—the coming and going of human figures—each has its proper place within nature's canvas.

In modernist architecture, space is constructed by inspecting lines of flow: the movement of people, light, and wind. Space is laid down piece by piece, often proceeding from the parts to the whole. In addition, studies of human settlements around the world have shown us that local environments inform which design considerations are most valued in a given place, leading to a diversity of architectural forms. In the most general sense, social and environmental conditioning makes genuine breakthroughs in architecture quite difficult.

However, much of MAD's work, apart from occasional hotels and private residences, are public buildings. Through these buildings, Ma Yansong's designs seem to pull apart the assumptions of modern architecture. He utilizes membranes to design worlds in which "inner" and "outer" are in constant exchange with one another; where contact between the individual and the outside world facilitates the infusion of interior and exterior space. In his work, the architecture itself becomes part of nature. Modern building techniques, with steel frames and glass walls, can only grow as tall or large as gravity allows. MAD's work is defined by an effort to escape this limitation through a transcendence of time and space.

Looking back on "Fish Tank," one of the earliest works produced by MAD, we are struck by the image of goldfish swimming freely within an involuted membrane, as if liberated from gravity. Today, this fish tank, a symbol of MAD's beginnings, still sits in their offices, occupying one large corner of a ping-pong table, smiling enigmatically.

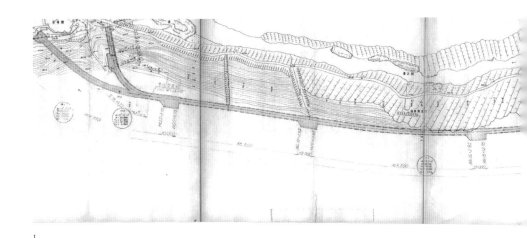

1

1. 1996年清津峡トンネルオリジナル施工図　写真提供：十日町市   1. Construction plan of the original tunnel from 1996, courtesy of Tokamachi City

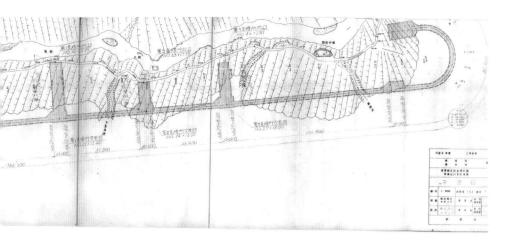

光のトンネル　　　　　　　　　　　The Tunnel of Light

私が北川フラム氏に初めて会ったのは、2016年に「瀬戸内国際芸術祭」のシンポジウムに参加した際であった。かねてから、北川氏がアートによる農村部や離島などの地域活性化に力を注いでおられることは聞き及んでおり、そのたゆまぬ努力に敬服していた。その後2018年に開催される第7回「大地の芸術祭 越後妻有アートトリエンナーレ2018」に私たちMADアーキテクツも招待を受け参加する運びとなり、十日町市の清津峡トンネルをリニューアルすることとなった。

——馬岩松

I first met Mr. Fram Kitagawa at a conference discussion at the 2016 Setouchi Triennale. By then, I had known about his efforts in revitalizing the countryside and admired his persistence. Later, he invited us to transform the Kiyotsu Gorge Tunnel at Tokamachi as part of the 7th Echigo-Tsumari Art Triennale, 2018.

— Ma Yansong

川端康成は『雪国』の冒頭で、「国境の長いトンネルを抜けると雪国であった」と冬の上信越地方の光景を描いているが、清津峡はまさにその通りの風景であった。

越後妻有は有数の豪雪地帯だが、中でも清津峡は最も積雪が激しく、一年のうちの長い期間大雪に覆われる。日本最長の信濃川がもたらす肥沃な土壌と大陸からの季節風によって、稲作に適した土地である。しかしながら、険しい山の地形ゆえ、土地を整備し棚田を開墾せざるを得ない。山々、信濃川、大雪が相まって、越後妻有独特の風景と文化が育まれてきた。

ここで暮らしてきた人々にとっては、変化のはっきりした四季、田んぼ、村落、家屋、農耕用具などは、代々伝わってきた、見慣れた風景だという。果てしなく長い雪の季節は、春の到来と客人の訪れをじっと待つ。

しかし、現代化そして都市化とともに、若い世代は都市部に流れ、次第に農業離れが進み、この地で生活するという根底が崩れ始め、経済は衰退の一途をたどった。その上、まさに北川氏の言うとおり、「世界で一、二を争う豪雪地帯で米を生産することが、そもそも非効率だと考えられるようになった」。

北川氏は、アートが苦境を突破するきっかけになることを願い、その指揮のもと「人間は自然に内包される」が大地の芸術祭・越後妻有アートトリエンナーレの一貫したテーマになっている。彼はアーティストたちに、日常の出来事や光景を、作品を通して表現するようインスピレーションを与えた。清津峡トンネルのリニューアルにあたっては、MADアーキテクツに対する要望として、自然と人間や文明のつながりを表現してほしい、とのことだったので、これを念頭に「Tunnel of Light」の構想を作り上げた。

清津峡は日本三大峡谷のひとつで、独特な地形をしており、両側の岩壁は切り立ち、渓流がくねくねと伸び、そこに流れるのは冷たく澄んだ雪解け水である。岩壁の片側にはかつて登山道が設けられ、登山者や観光客が絶え間なく訪れていた。観光業はこの地域の新たな産業となった。

"The train came out of the long tunnel into the snow country," writes Yasunari Kawabata. A depiction of winter in the Joshin-Etsu region, this opening line of *Snow Country* is also a fitting description of winter at Kiyotsu Gorge.

In Echigo-Tsumari, both culture and landscape are shaped by mountains, rivers, and heavy snow. Within the region, Kiyotsu Gorge has the densest winter snow, and is therefore closed to visitors for most of the year. Rice cultivation and paddy fields are another hallmark of the area, where farmers once cultivated rice in terrace fields on the mountainous terrain. The fertile soil contained essential minerals brought by the Shinano River, the longest river in Japan, while continental monsoons further contributed to the cultivation.

For those who grew up here, the vibrant seasons, paddy fields, farm tools, houses, and villages have always been part of a life that has continued for generations. In long winter seasons when the region was closed due to heavy snow, people waited for spring to come, when warm weather would bring visitors and guests.

Yet, things began to change. Younger generations began moving to cities, leaving the farming lands behind. Both local life and local economies crumbled in the shadow of modernization and urbanization. As Fram Kitagawa once said, "to grow rice in a region with a world-renowned level of snow has been considered extremely inefficient."

Perhaps art could lead the local economy out of the impasse, thought Fram Kitagawa. For that reason, the ongoing theme of the Echigo-Tsumari Art Triennale has centered around the topic of "how humans should live with nature." He encourages artists to engage emotionally with ordinary objects and scenarios, and use art to express feelings and emotions. Similarly, the project undertaken by MAD Architects in the area should highlight the relationship between nature,

しかし、落石事故が起こったことで、突然終わりを迎えてしまう。1988年、落石により一人の登山者が命を奪われ、遺族が管理不全として中里村（当時、現十日町市）を提訴、中里村は敗訴した。清津峡登山道は閉鎖され、観光客は来なくなり、地元経済は急激に落ち込んだ。

登山道閉鎖は、地元の人たちの心に重くのしかかった。単に経済的な打撃からだけではなく、清津峡登山道は彼らにとって非常に大切な、自然を楽しむことができる場所だったからだ。登山道が閉鎖されて8年、人々の記憶の中にある、雄大で壮麗な自然美は二度とたどり着けない彼方になってしまっていた。環境省、文化庁、新潟市など、多方面による調整の末、1996年ついに登山道に代わってトンネルが完成し、人々は峡谷の奥にある見晴所まで行って清津峡の景色を楽しむことができるようになった。トンネル完成後、観光客が来るようになり、清津峡周辺にはホテルなど商業施設ができた。また、このあたりの温泉は200年以上の歴史があり、パイプを整備して温泉を引き、住民や観光客が利用できるようにして、次第に温泉街ができた。しかしながら、コンクリート造りの単調なトンネル内は退屈で、やがて人々の足は遠のいた。トンネル開業からほどなく、清津峡や周辺地域は再び寂れてしまう。1996年の開業から2018年のリニューアルまでの22年間、トンネルは雪解け水の傍らで静かに新生のときを待っていた。

humans, and civilization. To continue this spirit, MAD proposed the "Tunnel of Light."

The Kiyotsu Gorge is one of the three major gorges in Japan, with a unique landscape holding a river formed by meltwater, and towering rocky cliffs of a columnar jointing structure. On one side of the cliffs, a hiking trail was once built to attract more visitors to this dramatic view of nature. The valley welcomed a constant stream of hikers, and tourism soon became the new economic pillar.

In 1988 (Showa 63), a rockfall took the life of a hiker. The hiker's family sued the Tokamachi city hall for management negligence, and won. The tragic incident ultimately put an end to local tourism, and entering the gorge was prohibited without exception. The local economy once again plummeted.

The prohibition not only hit the tourist trade hard; the local community was also affected. For them, the gorge had been part of everyday life. Now, it all became an irretrievable memory. The closure lasted for eight years until 1996 (Heisei 8), when the Kiyotsu Gorge Tunnel was completed. The project was a joint effort between the Niigata prefecture, the Environment Agency, and the Agency of Cultural Affairs. The tunnel branches into viewing platforms located deep within the gorge, once again allowing visitors access to the unique scenery of Kiyotsu Gorge. Soon, the flood of visitors resumed. Restaurants and hotels opened. Hot springs with a history of more than 200 years were channeled from the mountains into houses and hotels, which gave the place its name; the Onsen Street. Yet, the experience of the concrete tunnel paled in comparison with that of the original hiking trail. Not long after its initial popularity, business around the gorge and villages became stagnant. From 1996, until its renovation in 2018, the Kiyotsu Gorge Tunnel sat idly beside the river, waiting for a new beginning.

2

2. エントランス施設「潜望鏡」リニューアル前 ©MAD Architects　　　2. Before the renovation of the entrance facility "Periscope," ©MAD Architects

3

3. エントランス施設「潜望鏡」リニューアル後、施設よりトンネル入口を望む 撮影：馬岩松

3. After the renovation of the entrance facility "Periscope," and the view of the tunnel entrance from the facility. Photography by Ma Yansong

4

5

4. 上部、反射鏡を通して外部の風景を内部に引き入れる　撮影：馬岩松

5. 建物正面　撮影：Nacasa & Partners Inc.

4. The roof's ridge creates exterior views through height differences and mirrors. Photography by Ma Yansong

5. Building elevation, photography by Nacasa & Partners Inc.

プロローグ

駐車場で車を降り、清津峡の川の流れに誘われるように道を進むと見えてくるエントランス施設の切妻屋根。その屋根の急勾配は、厳しい冬の雪を想起させる。

施設は二階建てで、一階はチケット売り場、カフェ、土産物店を兼ねた場所である。地元住民による精巧な手工芸品が並び、開放的で温かく、観光客の休憩スポットとなっている。

二階はユニークな設計で、当地の温泉を引いた足湯空間として、五感で感じるトンネル体験のプロローグとなっている。空間内部は円錐形で、天井の開口部には反射鏡が設置されている。緻密なシミュレーションを経て、外を流れる清津峡渓流の様子が、反射鏡を通して空間内部に映し出される仕組みになっている。まるい足湯の上に、温泉の湯気が立ち込め、視覚と嗅覚が曖昧に重なり合う。来場者はここで休憩し、両足を温泉に浸しながら、頭上を仰ぎ見ると、清津峡渓流が逆さまに目に映り、耳には渓流が流れる音と鳥のさえずりが降り注ぐ。――これらすべてが、《Tunnel of Light》の旅で出会う光景を暗示する。

清津峡は国の指定する名勝天然記念物であり、国立公園の一部に指定され、ここで行われるあらゆる建築は事の大小に関わらず、全て環境省及び文化庁の規定する厳格な法令や規則をクリアしなければならない。――屋根のかたち、建築材料、色彩など、全て当地の伝統建築にそぐうように非常に厳格な規制があり、ひとつひとつの設計段階ごとに、まずは十日町市の認可を取得したうえで、具体的な設計に入っていくという、非常に複雑で緻密な作業であった。

Prologue

The trail of visitors begins at the parking lot. As if guided by the gurgling river nearby, people stroll along a narrow path and arrive at the entrance building; its steep roof a reminder of the area's heavy winter snow.

The building has two floors. Ticket offices, a café, and a souvenir shop selling fine handmade crafts made by the villagers occupy the ground floor. The space emits a warm and open atmosphere; a place to rest before people set out on their journey.

As a prequel to the tunnel, an ingeniously designed foot spa makes up the upper floor. Above, the top of the conical roof forms an unusual skylight framed by prismatic mirrors, capturing the running river in the distance as an inverted reflection. The space is shrouded in mists coming from the bath, filled with aromatic and visual ambiguities. When visitors immerse their feet in the water channeled from the natural hot spring, and look up, they see the reflection of the river drifting across the sky. As the gurgling and chirping sinks in, the journey of the Tunnel of Light begins.

Kiyotsu Gorge is also a nature conservation area. Architecture must therefore comply with the meticulous regulations of the Environment Agency and the Agency of Cultural Affairs. The height, roof, materials, and colors must meet these requirements and follow local architectural styles and tradition. Every phase of the project required approval from the Tokamachi City government before further progression.

6

6. 写真提供：十日町市観光協会　　　　　　　　6. Courtesy of Tokamachi Tourist Association

7

8

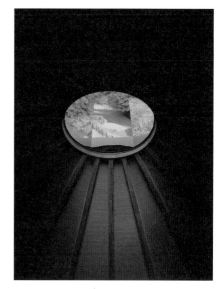

9

7. 二階の足湯空間　撮影：Nacasa & Partners Inc.

8. 異なる角度からは、異なる風景が見える　撮影：Nacasa & Partners Inc.

9. 鏡の中に清津峡が逆さまに流れる　撮影：Nacasa & Partners Inc.

7. Foot-bathing space on the second floor, photography by Nacasa & Partners Inc.

8. Different views from different perspectives, photography by Nacasa & Partners Inc.

9. The reflection of the Kiyotsu Gorge flowing in the mirror, photography by Nacasa & Partners Inc.

10

10. 一階の受付と休憩スペース　撮影：Nacasa & Partners Inc.

11. 天井開口部が反射鏡となり、外部の風景を映し出す
撮影：Nacasa & Partners Inc.

10. Service and lounge space on the ground floor, photography by Nacasa & Partners Inc.

11. The skylight becomes a periscope, reflecting the view outside. Photography by Nacasa & Partners Inc.

11

12

12. 頂上部の鏡が足元の温泉に映し出す逆さまの風景がぼんやり見える ©MAD Architects

12. The reflection of the mirror above your head can be faintly seen in the hot spring pool below your feet. ©MAD Architects

当地は日本屈指の豪雪地帯であり、エントランス施設には雪対策が必須であった。屋根は積雪防止のため、勾配の急な切妻屋根を採用する必要があった。敷地が国立公園内に位置するために高さの制限もあったが、MADはこの難しい屋根形状の条件を活用しある仕掛けを導入することにした。屋根のてっぺんの開口部に反射鏡を設置し、外の清津峡の風景を上下逆さまに室内に投影する仕組みである。

Due to its location in one of the strongest snowfall regions in Japan, the entrance building must be resilient against harsh and heavy snows. MAD therefore designed the building with an A-shaped roof that allows snow to slide off with ease. Although regulations strictly limit the building's height, MAD's design incorporates some unique features. The hipped end consists of sections overlapping each other at different angles, slanting away from the elevation before creating an opening at the top. Here, prismatic mirrors are installed within the open skylight to direct reflections of the surrounding landscape inside.

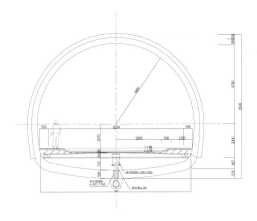

パノラマステーション断面図 S=4:50

13

14

13-15. 鏡の中の景色は季節の変化によって変わる
撮影：Nacasa & Partners Inc.

13-15. The view in the mirror varies with the changing of the seasons.
Photography by Nacasa & Partners Inc.

15

16

17

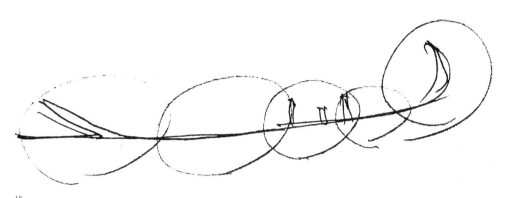

18

16. 入口付近　撮影：馬岩松

17. 第四見晴所（パノラマステーション）付近　撮影：馬岩松

18. 各ゾーンで異なる光は見晴所ごとに異なるテーマと関連している

16. Close to the entrance, photography by Ma Yansong

17. Close to the fourth viewing platform, photography by Ma Yansong

18. Light will change in different areas depending on the theme of the viewing platform.

トンネル

エントランス施設から歩くとほどなく、トンネルの入口に到着する。入口を入るとトンネルは左に大きくカーブし、そこから一本の長い通路が見える。外部の光が届かない空間で、ぼんやりした照明の光が目に入ってくる。

トンネルの全長は750メートルで、4つの見晴所から清津峡を望むことができる。仄暗い照明の光はトンネル内の各ゾーンによって色が変化し、その色は各見晴所のデザインと関連している。来場者は光の変化とともに感覚や感情の変化を体験する。

初めの明るい黄色の光から緑色の光に変わる辺りで、第1見晴所に到着する。MADはこの第1見晴所を、何も手を加えず元々の状態のまま残すことにし、来場者に清津峡トンネルの建設当初の姿を体験してもらえるようにした。ここからは峡谷の柱状節理を眺めることができ、眼下を流れる清津川の絶景を味わうことができる。第1見晴所には日本の桜のような淡い白色の光を採用した。

さらに歩いていくと、鮮やかなオレンジ色の光が、第2見晴所と第3見晴所の未来的な空間とマッチして、まるで異星にいるかのような不思議な高揚感をもたらす。ゆっくりと最後の見晴所に近づくにつれ、青色の光が気分を落ち着かせる。このように光の色の変化が、トンネルの単調な一本道に風景や体験をもたらし、感覚や感情の変化を引き起こすことで、「通路」と「目的地」の境界を曖昧にさせる。トンネルの空間全体をひとつの体験として融合させるのである。

トンネル内には音が響いている。歌でもなく、音楽でもなく、どこの言語とも分からない、誰かが口ずさんでいるような声が、人々をトンネルの奥へと誘う。

The Tunnel

After a few minutes' walk from the entrance facility along the hillside, one arrives at the entrance to the tunnel. Inside, after a sudden turn to the left, a seemingly endless passageway reveals itself. Daylight disappears unseen. An ungraspable warm luminescence takes over.

The 750-meters-long tunnel is the only visiting route, along which four viewing platforms provide visitors with intimate views of the splendid scenery in Kiyotsu Gorge. The main branch of the tunnel is lit with different colors to reflect the changing moods and rhythms at each platform, adding extra layers to the sensory experience.

As the lighting changes from bright yellow to light green, visitors arrive at the first platform. Here, MAD has preserved the historic state of the tunnel. From the first platform, visitors look out upon the vivacity of nature, where columnar joints make up the surface of the cliff, and jade-like water rushes along the base of the gorge and tumbles down the mountain. Appropriately, this section is lit in cherry blossom pink, a symbolic color of Japan.

The second and third platforms - futuristic and eerie – are lit in an exhilarating tangerine. As one walks away from the otherworldly, and nears the end of the journey, the lighting stealthily cools down. The visitors slow their pace in this calm blue. While walking through the tunnel was once simply a means to an end, the changing colors transform the bland interior into an active part of the scenery. The use of light engages with visitors' emotions and contributes to a wholistic visiting experience. The boundary separating ways and destinations is blurred; the experience uniting divided spaces as one.

In addition to the remarkable lighting, visitors are also surrounded by sounds. The murmurous flow of nature, as if humming obliviously in a language unknown to man, lures visitors into its depth.

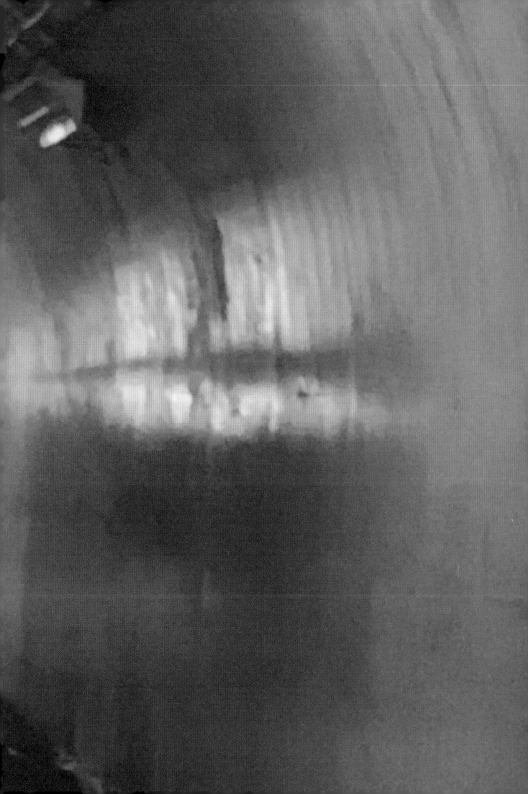

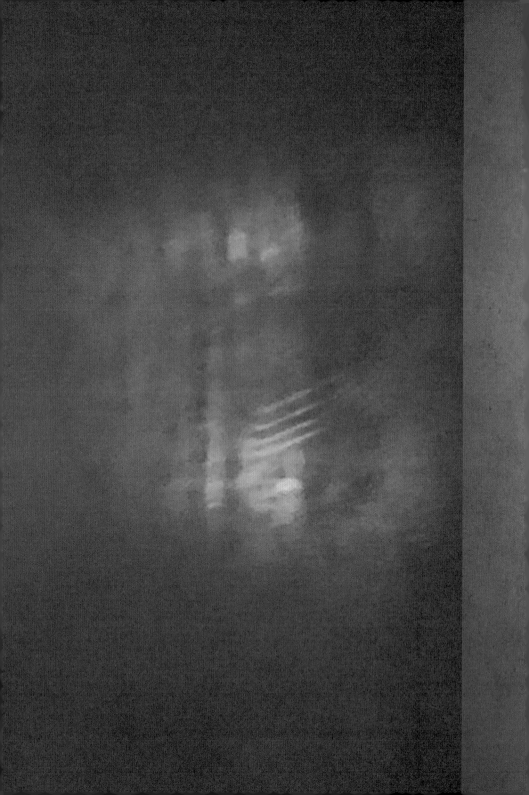

19. ［P.33-36］それぞれの光の色は各見晴所と呼応し、トンネルと見晴所の境界を曖昧にする 撮影：馬岩松

20. ［P.37-44］撮影：Nacasa & Partners Inc.

19. [P.33-36] Various lighting colors echo the viewing platforms and entrances, while blurring the boundary between the tunnel and the platforms. Photography by Ma Yansong

20. [P.37-44] Photography by Nacasa & Partners Inc.

トイレ

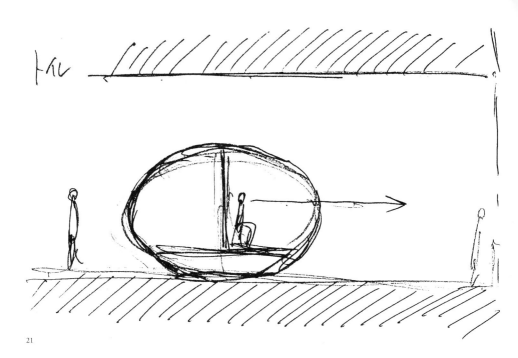

21

21. 来場者は見晴所の開口部に立ち、眼下の清津峡の景色を見ている。
この光景は「見えない泡」（トイレ）の中にいる人の目の前にある風景にも
なる

21. Visitors stand at the outer edge of the viewing platform and look down to
enjoy the view of Kiyotsu Gorge. This act itself forms the "scenery" in the eyes
of people within the "Bubble" bathroom.

見えない泡

第1見晴所を過ぎてさらに奥へ進むと、第2見晴所に到着する。

ここには未来的な鏡面に覆われた「見えない泡」が置かれ、トンネルの灰色の内壁と峡谷の美しい風景を反射している。まるで未来からやって来たカプセルのようで、しかしその姿ははっきりとは捉えられない。

この「見えない泡」は実はトイレである。峡谷側に向いている方は誰でも利用できるトイレ、反対側は女性用トイレである。表面がマジックミラーになっており、外側からは鏡のように反射し内側を見ることはできず、一方、内側からは外側が透けて見える。トイレを利用する際、内側から外の清津峡の美しい風景はもちろんのこと、見晴所を行き交う人々の姿まで見える、という特別な体験ができるのである。

内側からは風景が見えるが、風景を見ている人々からはあなたを見ることはできない。

The Bubble

Moving forward, one soon arrives at the second platform. Here, visitors encounter a "bubble" reminiscent of a spacecraft that can disappear at will. The Bubble has a mirror surface, reflecting both the grey interior and the natural scenery outside, making itself vanish from sight.

Functionally, the Bubble serves as a restroom. The side closer to the valley is unisex, while the other side is for women only. The outer layer of the Bubble uses a one-way see-through glass material. When seen from within, the wall of the Bubble becomes a full-length window, while when seen from outside, it becomes a reflective mirror. Visitors inside the restroom can continue to enjoy the view of nature, which now includes other visitors strolling around.

You watch the scenery from inside; the scenery watchers see you not.

23

22-25. 撮影：Nacasa & Partners Inc.   22-25. Photography by Nacasa & Partners Inc.

24

25

新生版「見えない泡」

第8回大地の芸術祭越後妻有アートトリエンナーレ開催に際して(当初予定では2021年開幕)、MADは北川氏より招待を受け、「見えない泡」のある空間をさらに改造し、3年前の作品をさらにリニューアルすることとなった。

春になって雪解け水が新たに川に流れ込むように、MADはリニューアルの新たな設計において、清津川が勢いよく流れ込むような、躍動感とエネルギーに満ちた空間を創り出そうと考えた。

白と黒の螺旋に覆われた空間に一歩踏み入れると、時空のトンネルに入り込んだような錯覚にも、渓流の渦に飲まれるような感覚にも陥る。外の清津川が、白と黒の螺旋を反射した「見えない泡」の鏡面を介してトンネル内部に流れ込む。その視覚効果で、人が動くとともにまるで万華鏡の中に入り込んだかのように、目に映るものがクルクルと変化する。足を止めると「見えない泡」の表面は空間に溶け込み、視覚からかたちが消える。トンネル外部の渓流の音が響きわたり、五感が立体的に強烈に刺激される。

Rebirth of the Bubble

The Bubble was initially constructed in 2018. In the spring of 2021, MAD was invited by Mr. Kitagawa to reinvent the space around the Bubble for the 8th Echigo-Tsumari Art Triennale, which was originally scheduled to open in 2021.

The river's nascent meltwater signals the coming of spring. MAD's proposal attempts to abstract and capture the exuberant spirit of the Kiyotsu River, providing visitors with an immersive and dynamic spatial experience.

Being inside the space, covered in black and white helices, one feels as if they are passing through a warped time-space, or falling into the many whirlpools of the lively Kiyotsu River. The energy of the river now floods into the tunnel through the geometric pattern flowing around and onto the Bubble. As visitors walk around the platform, what they see also moves, similar to a kaleidoscope. When they stand still, the movement also comes into a halt, and the form of the Bubble vanishes into the surrounding environment. The gurgling outside endlessly pulsates and fills the space with a complete stimulation of the senses.

26

26. 写真提供：十日町市観光協会

26. Courtesy of Tokamachi Tourist Association

27

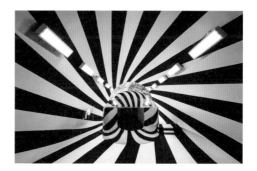

27. 2021 年 MAD が「見えない泡」の空間に新たに設計した「Flow」
撮影：中村侑

27. In 2021, MAD created a new work within the "Bubble" space, titled "Flow."
Photography by Osamu Nakamura.

異世界

トンネルの第3見晴所に差し掛かると、突然無数の道がある異世界に迷い込んだように感じる。

第3見晴所の壁面には、不規則に並んだしずくのような鏡面が散りばめられ、温かみのあるオレンジ色の光を発している。鏡面を覗き込むと景色は反射しながら延々と連なり、未知の空間へつながっているような感覚に陥る。

見晴所の中を歩くと、トンネル形状にそってさまざまな角度を持つ鏡面に、自らや他の人々の姿が何重にも反射し複雑な景色を映し出す。

MADはもともと使われていた照明と配線を利用して、しずく型の鏡面と組み合わせ、オレンジ色の間接照明とした。ランダムに見えて実は緻密に計算して配置され、それぞれ微妙に角度の異なる鏡面は、無数の偶然性と可能性に満ちた場景を創り出す。

峡谷への開口部に向かっていくにつれ、人の動きは四方八方にある「目」に捉えられ、徐々に鏡面に反射した人の姿は青々とした大自然に溶け込んでゆく。鏡面に散りばめられた自然はオレンジ色のバックライトに照らされ、昆虫の目に映る不思議な宇宙のようでもある。

An Eerie Space

At the fork in the tunnel, one takes a turn and enters the third platform. Immediately, one falls into a junction of the otherworldly, an intersection that leads to countless passageways.

Irregular mirrors shaped like water drops emit a warm tangerine light around their peripheries. As visitors linger, these mirrors become eye-like. Here on the arching concrete wall of the tunnel, they become windows that lead to unknown worlds.

Visitors see the fractured images of themselves and the other visitors in these mirrors. Through multiple reflections, the human figures become whimsical and absurd.

The installation of the indirect tangerine lighting utilized existing lighting fixtures for convenience. Each lighting unit and drop-shaped mirror form a modular unit. What appears to be a random composition was in-fact carefully configured, with mirrors placed at calculated angles to create a space filled with contingencies.

As one walks to the end of the platform, reflections of nature begin to blend with the distorted human figures. Now, the rich tangerine lights of the mirror encircle a fractured green; a myriad of worlds seen through compound eyes.

28-29. 見晴所の開口部に向かっていくにつれて、鏡面に映る光景はトンネル内の暗いオレンジ色から峡谷の青々とした大自然へ変化していく
撮影：Nacasa & Partners Inc.

28-29. As you walk towards the outer edge of the viewing platform, the view in the mirror gradually changes from the dark red of the tunnel to the verdant nature of the canyon. Photography by Nacasa & Partners Inc.

30

31

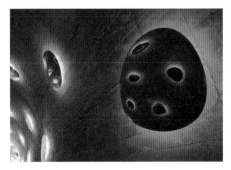

32

イリュージョン

青い光に沿って歩いていくと、外部から差し込む光が徐々に見え、さらに進むと第4見晴所（パノラマステーション）——「ライトケーブ（光の洞窟）」に到着する。

それまでの3つの見晴所が峡谷に対し横から岩壁を見るのに対し、唯一このパノラマステーションは峡谷に正対して見ることができる。見晴所自体に奥行きがあり、外の切り立った崖の遥か下に清津川が流れるため、パノラマステーションの入口では清津川を実際に見ることなく、激しい川の流れの音を聞きながら空間を進む。

MADは床面を水深の浅い水盤とした。外部の実際の景色と水面に逆さに映る景色とが融合し、水面でひとつの円形状の風景となるようにした。

トンネルの壁面も、鈍い反射のステンレスパネルで覆うことにより、峡谷の切り立った崖面とその下に流れる清津川の風景をゆがめながら壁面と天井に鈍く映すことになる。自分の立つ場所によって、天井にうっすらと流れる川の姿を確認することができる。トンネル端部まで張られたステンレスパネルによって、外部と内部の境界線が曖昧になり、外部が内部に浸透するような効果を醸し出す。峡谷の絶景がトンネルに映り込み、自然と人間、そして人工物と融合し、新たな作品となる。

「パノラマステーション」からは、空は逆三角形に切り取られたように見える。下を流れる清津川は勢いよく流れ込み、足元に消える。空や峡谷を流れる川はさまざまに形状を変えながら絶え間なくトンネル内部の空間に反射し、清津峡の動と静の相まった美しい光景を目に焼き付ける。水盤の中を進むにつれて、川の流れる音は次第に大きくなり、それと同時に、峡谷の底を流れる清津川の流れが目に飛び込んでくる。

「ライトケーブ」の水は清津川から引いていて、来場者は靴を脱いで水盤に入ると、炎天下の夏日でもそ

Scenery Beyond Reality

As the journey approaches its end, the lighting changes into a hushed blue. Daylight trickles through the tunnel as visitors walk straight ahead to the fourth platform - the light cave.

Unlike the first three platforms that open onto the opposite cliff, the fourth platform looks out over the winding stretch of the gorge. Because the space itself has a greater depth, the river lying far below the cliffs is hidden from the view when people first enter. The sounds of the unseen river rise from below and lure the curious visitor to come closer to nature.

MAD designed a shallow pond that stretches all the way to the end of the platform. The reflective water surface brings the outside scenery into the tunnel. The water almost brims over the edge of the platform and seamlessly joins the actual landscape, forming an infinite journey.

Inside the arching tunnel, sheets of mirrored, polished stainless-steel line the inner wall through to the end of the cave. The low reflectivity of the stainless-steel sheets makes the soft reflections of the rocky cliff and river indistinguishable from each other, with visitors uncovering vague clues of the flowing river in the metallic surface. As the visitor moves, so too do the reflections. The edge of the tunnel is where the reflections meet the source. Through the abstractions, the beautiful scenery oozes into the monotonic semicircular space. Nature, human, and materiality are fused into a new form.

Standing on the fourth platform, one sees the piercing cliffs cut the sky into an inverted triangle. Far beneath one's feet, the Kiyotsu River rushes towards and past them. The bewildering, dynamic reflection of the sky and river channels natural beauty onto the platform, from where visitors enjoy an unimpeded view of the gorge. As one walks across the shallow pond and

の冷たさを感じることができる。人が来ると、水面に波紋が広がり、風景は揺らぐ。人が居ない時には水面は静まりかえる。動と静が風景に新たな要素を加え、人々の行動によって水面に投影される風景が変化する。──同じように感じる時間においても、自然は常に静止しているわけではなく、人々の行動、往来も、自然にとってみれば、つかの間の出来事である。

開口部に近づくにつれて、トンネルへと吹き込む風を身体に感じ、季節の変化とともに日々表情を変える清津峡の自然の風景を捉え、来場者はそれぞれの心に、その時感じた感情と共にこの風景を刻み込んでいくことになるだろう。

MADは、光や風、温泉水や川の水、反射や色と、極めて基本的な要素のみを使用しつつ、清津峡に注意深く寄り添いながら、それぞれの見晴所の特性に加味し、全体としてひとつの「旅」を紡ぎだした。

approaches the outer edge, the Kiyotsu River reveals itself through the sound of rushing water becoming louder and louder.

The shallow pond in the "light cave" is channeled from the Kiyotsu River. Visitors can take off their shoes and walk through the pond, offering soothing relief from the summer heat. Movements in the pond ripple the water, shattering the reflection before it gently returns to its natural stillness. In essence, the water translates human movements into a unique language that communicates with the surrounding nature. A supposedly still surface is activated by movements, uncovering a new medium for experiencing nature. Nature is not as still as it might seem; the time taken to pass through the tunnel is only a fleeting moment in the history of Kiyotsu Gorge.

Near the opening, wind touches visitors more tenderly. With it, the wind carries the nuances and expressions of the different seasons flowing through Kiyotsu Gorge, leaving people with stirring emotions.

Light, wind, hot springs, rivers, reflection, and colors; the choice of design elements are minimal. The design by MAD thoughtfully incorporates these essential elements into unique spaces on the platforms and around the gorge. The revitalization of the tunnel weaves together a wholistic, sensory journey.

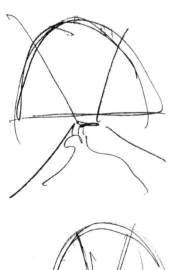

35. 撮影：Nacasa & Partners Inc.

35. Photography by Nacasa & Partners Inc.

38

37. 風景は来場者の行為にともなって変化する。誰もいない時、あるいは
人の動きが止まっている時、水面は風景を鮮明に反射する　撮影：中村侑

38. ライトケーブ（光の洞窟）は唯一峡谷に正対する空間である。峡谷に
逆三角形に切り取られたような空とその下に流れる清津川が、パノラマス
テーションに足を踏み入れると目に飛び込んでくる
撮影：Nacasa & Partners Inc.

37. The landscape view changes from moment to moment depending on
the visitors' movements. When there is no visitor, or when the crowd is
still, the water surface can reflect extremely clear images. Photography by
Osamu Nakamura.

38. The "Light Cave" is the only viewing spot directly facing the gorge. From
here, visitors see the sky cut into an inverted triangle by the canyon and Kiyotsu
River flowing underneath it. Photography by Nacasa & Partners Inc.

39

設計記述

「新しい風景」
早野洋介（MADアーキテクツ共同主宰）

当初、十日町市から大地の芸術祭を通して提示されたプロジェクトの内容は以下の3つであった。

1　トンネルの入り口脇の建物が来坑者のための待合場として利用されているが、老朽のため これを建て替えて、現在トンネル内部に設置されている管理オフィスやチケット売り場を移設し、更にショップやカフェを併設したエントランス施設の設計。

2　トイレの併設されている駐車場からトンネル入り口まではかなり距離があり、更にトンネル内部にもトイレが無いために、来坑者及び近隣住民からのトンネル内部にトイレを設置して欲しいという要望に応えるため、第1見晴所にトイレの設置。

3　一番奥の見晴所（パノラマステーション）の空間をアート的な手法でイメージを変えて、より多くの人が訪れるような場所にして欲しい。

第7回大地の芸術祭（2018年）に向けて、マ・ヤンソンと二人で始めて現地を訪れたのが2016年、年の暮れも迫った12月末。冬には積雪4mを超すという豪雪地として名高い十日町。「今年は雪が少ない」と聞いていたが、それでも初めて訪れる者からすれば、まさに「雪国」の世界。風景を白く染めた雪でさまざまな音が消えてしまったような感覚を覚えながら、集落の一番奥に位置するトンネルに到着し、視察を行う。

芸術祭関係の他のプロジェクトと違い、清津峡トンネルは市が管理するため、年が明けてすぐの1月中旬に十日町市長へのプレゼンをすることになる。その際に私たちが提示したのは、上記の3つのみを行うのでは、この清津峡のイメージを根本的に変換することは難しいだろう。やるのであれば、エントラ

Design Process

A New Scenery
Yosuke Hayano, MAD Architects Principal
Partner

As part of the Triennale, the renovation project initially had to fulfil the following three requirements made by the Tokamachi city hall:

1. The facility at the entrance of the tunnel had historically served as a resting place for visitors. Having aged significantly, the facility required rebuilding in order to effectively maintain its role. The managerial and ticketing offices which had previously been housed inside the tunnel were also to move into the entrance facility, along with new shops and a café.

2. The old tunnel did not have a restroom, with visitors previously having to walk a great distance to the restroom at the parking lot. Visitors and residents had raised this issue and, therefore, the renovation aimed to offer a restroom on the first viewing platform.

3. The fourth platform required renovation with an artistic approach to create a highlight for the journey and to attract more visitors.

At the end of 2016, Ma Yansong and I traveled to Tokamachi to investigate the site in-person. Tokamachi is a place famous for its snow, which sometimes builds to a depth of four meters. Before embarking, I had heard that the approaching winter would see less snow. However, when we visited for the first time, our experience was comparable to the world depicted in Snow Country. Crossing a hushed and open landscape covered in a white blanket - a scenery unique to places with heavy snow - we arrived at the entrance of the Kiyotsu Gorge Tunnel located far into the village.

Unlike other projects in the Triennale, the tunnel

ンス施設、トンネル部分、そして4つあるすべての見晴所をそれぞれ異なった空間として、来坑者が清津峡の美しい自然を体感できる場とし、それらを連続して経験することで、全体としてアートという新しいフィルターを通して清津峡の自然に触れる「旅」にするというコンセプト。

通常「トンネル」というのは、物理的に繋がれていない2か所を、穴を掘ることで繋げる交通インフラ的な意味合いが強いが、この清津峡トンネルはそうではなく、このトンネルがなければ見ることのできない自然の風景と人とを繋げている。それはとても不思議な魅力であり、別の角度から見れば、深海の奥底に潜んでいる潜水艦が、細い潜望鏡を水面に出し外界の様子を眺めるように、このトンネルも堅い岩の中に潜みながらも、4か所の見晴所を伸ばすことで、清津峡の美しい風景に繋げていると考えられる。それならば、自然の異なった要素、光や水、風などを利用し、それぞれの見晴所で異なった感覚を刺激する風景を拡張することが、この清津峡トンネルの持っている魅力を最大化することに繋がるのではないか。

外部から見晴所という目的地に繋げるだけのトンネルではなく、日常から離れ、訪れたそれぞれの人の感性と自然の要素を通して、清津峡の新しい風景を訪ねるアートの旅。歩きながら見つけるのは自然の風景であるのと同時に、心の中に湧き上がる感情に耳を傾け、そしてまた入ってきた時とは少し変化をもって日常に帰っていく。そんな場所になって欲しい。そのような話をすると、市長も「非常に良いコンセプトだと思います。ぜひ思うようにやっていただければ。ただし、予算は限られているので、スマートに行ってください」とのことで、ここからいかにこのイメージを、さまざまな条件によって妥協することなく、実現させていくかという具体的な設計に入っていく。

その場で北川フラム氏からも助言されたことは、このような規模の大きなプロジェクトでは、当初のコンセプトを保ちながら最終的な作品にまで仕上げるのは非常に難しいということ。単体で設置することができるアートの作品と異なり、建築や土木の規模

project was overseen by the Tokamachi city hall. Therefore, we were required to report our plan to the city mayor in the coming January, 2017. We noted at the time that the resulting scheme would not significantly enhance peoples' existing impression of the place, if we were confined to the three requirements mentioned at the beginning. As we were determined on reinventing the tunnel, we agreed to treat the entrance, tunnel, and four platforms each as a unique space to reimagine the experience of visiting the gorge.

Our vision was to renovate each space individually while maintaining an overall continuity. Our goal was therefore to create an intimate experience of journeying through the natural wonder of Kiyotsu Gorge.

Our common understanding of a "tunnel" is a passageway for creating shorter, more convenient paths through an impassible object. Tunnels are often thought of as infrastructure, and nothing more. However, this is not the case with the Kiyotsu Gorge Tunnel. Here, the tunnel connects visitors with an endearing nature that is otherwise inaccessible. In this sense, the tunnel itself is unique and inspiring.

Viewed from another perspective, the tunnel resembles a submerged submarine with its periscope peaking above the ocean. This particular "periscope" is embedded in hard rocks among the cliffs, through which people can observe and connect with the nature of Kiyotsu Gorge.

We therefore sought to entice nature further inside the tunnel. We chose elements such as light, water, and wind to invigorate the space on each platform. By intensifying senses and heightening our perception of nature, we aimed to reactivate the spatial potential of the tunnel.

The tunnel is more than a passageway that leads to various platforms. Instead, it fuses these "natural elements" with the sensibilities of bodies and minds,

になると、さまざまな条件や制限がかかり、その中で調整や妥協を強いられることで、当初のコンセプトの輝きが失われることをいかに避けながら実現していくかが問われてくる。

幸いなことに、このプレゼンテーションに出席されていたのが、プロジェクトの終わりまで関わることになる各部署の関係者で、コンセプトの内容及び、自治体のリーダーである市長の意向も皆で共有することができ、後はそれをいかに実現していくか、プロセスの中で何度もコンセプトに立ち返り、何を保持することが大切なのかという話し合いを重ねることができ、思った以上に当初のイメージ通りの姿で竣工を迎えることができたプロジェクトとなった。

これは清津峡トンネルが位置する清津峡温泉街の地元住民への説明会でも同様で、清津峡トンネルがもたらした賑わいと共に生活をしてこられた温泉郷の方々の中には、「芸術祭に伴ってトンネルの改修をする」という内容を聞いた際には反発の意見もあったと聞いたが、何度か機会をいただき、清津峡及びこのトンネルの歴史を踏まえ、その魅力をさらに向上させるためにどのような考えをもって設計を考えているかという説明会をさせていただき、さまざまなお話を聞かせていただいたり、議論をしながら、非常に良い形で地元のサポートも受けてプロジェクトを進めることができた。

技術的に非常に難しい条件として、この清津峡自体が自然保護地域に属しているために、環境庁および文化庁の定める条件の中で設計を行わなければならず、屋根の形状や色、素材など段階ごとに十日町市と共に確認を取りながら設計を進めた。

それと同時に、日本でも有数の豪雪地であるこの地で、設計するエントランス施設は、雪に対してどのような設計を行わなければいけないかという面で、協力会社であるグリーンシグマから多くのことを学びながら、コンセプトを実現することと、公共施設として長期にわたる使用に耐えうることを念頭に設計を詰めていくことになる。

recently escaped from the everyday hubbub to create a journey of art. As visitors stroll along the tunnel, senses and feelings are rediscovered. We hope to transform the tunnel into a place that, after people leave and return to the everyday, its subtle stirrings stay with them.

The mayor was highly supportive of the idea, saying "please, design it that way; but given that the budget is quite limited, you may need to employ ingenuity." Having received approval from the mayor, we began thinking about how to realize the concept within the constraints and move towards a tangible design.

Meanwhile, we also received advice from Mr. Kitagawa. He believed that maintaining the original idea for the entire duration of a project of this scale would be extremely challenging. Unlike individual artworks, this project involved both architecture and civil engineering, meaning many constraints would be taken into consideration during the design process. Negotiations and compromises would be unavoidable. As a result, the original idea could easily lose its shine. Realizing the idea without diminishing it became a major challenge for us.

Fortunately, all of the people who would later take part in the project attended our initial presentation, during which they became familiar with the design, the underlying idea, and the mayor's input. This universal understanding allowed us to work closely and cooperatively across all disciplines. Throughout the process, we continuously referred back to our initial idea when confronting questions such as "what is the most important aspect?" and "what should be included?" Therefore, it could be said that the final result is even closer to our initial idea than what we had expected.

In tandem with the scheme's construction, we delivered a series of presentations to the residents of Onsen Street, where the Kiyotsu Gorge Tunnel is located. People have lived on this street for generations, and have witnessed the short-lived prosperity brought by

駐車場で車を降り、清津峡の川の流れに誘われるように道を進むと見えてくるエントランス施設の切妻屋根。厳しい冬の雪を想起させる屋根勾配。その屋根の下の二階の空間では、かつてこの清津峡トンネルを掘削していた際に出たというこの地の温泉の湯を引き、中央部に足湯スペースを設けている。空間は円錐状になっており、その上部は外部に開放され、外の川の流れや鳥たちのさえずりが内部と外部を繋いでくれる。上部の開口部は大きな鏡で覆われており、そこには遠くに流れる清津峡の川の流れが上下さかさまになって映し出されている。訪れた人々は、足に大地の恵みを感じながら、空に流れる清津峡の姿を見ることで、これから始まる旅を感じることになる。

トンネルの入り口を入るとすぐに左に大きくカーブして、長い通路が見えてくる。壁面に設置してあった照明を部分的に外し、照度を落とすことで、洞窟の中を散策するような雰囲気を演出しながら、照明に特殊なカラーフィルムを被せることで、トンネル全体を5色の異なる色のゾーンに分け、それらが連なる色のトンネルとした。各色は各見晴所や入口に関連した色となっており、トンネルと見晴所と分けるのではなく、全体がシームレスな空間として繋がるようになっている。

トンネル中央部からトンネル全体に届くような音を流し、歌でもなければ曲でもなく、何の言語かも分からないが、確かに誰かが口ずさんでいるような音に誘われてトンネルの奥へと進んでいくことで、まるで時間を遡り進んでいるかのような雰囲気を醸し出す。

そしてたどり着く第1の見晴所は、当初の見晴所そのままの姿を残し、訪問者はこの清津峡トンネルの歴史、そして開口部から眺めることのできる柱状節理の荒々しい岩肌とその下を鮮やかな色を見せて流れる清津川の風景を眺めることができる。当初はこの第1見晴所に新設のトイレを設置する予定であったが、地元の方との話し合いを重ねる中で、清津峡トンネルの歴史を踏まえ、その上で今回の改修が行われたという時間の流れを来訪者の方にも感じながらそれぞれの見晴所を辿っていってもらったほうがよいのではと

the tunnel. Some objected to the project. We explained that the idea was to renovate based on the history of the gorge and tunnel; that the renovation would further enhance their appeal. Meanwhile, we also received many suggestions. These discussions helped us to gain support from the locals, who later helped us in pushing the project further.

As Kiyotsu Gorge is a natural heritage site, the design needed to be comply with parameters set by the Environment Agency and the Agency of Cultural Affairs. The shape, color, and materials of the roof required confirmation at each stage with the Tokamachi City government.

Meanwhile, we learned much from our partner company, Green Sigma, about how to design an entrance facility that could withstand the snow in one of the heaviest snowfall regions of Japan. The design was developed not only to realize the concept, but also to ensure the building's long-term durable use as a public facility.

After leaving the parking lot, lured by the flow of the river, the gabled roof of the entrance facility comes into view as visitors walk along the road, its steep sloping roof serving as a reminder of the heavy winter snow. Stepping inside, and moving to the second floor of the facility, a footbath area sits in the center with water sourced from the local hot spring. The upper part of the space is open to the outside, with the sounds of the flowing river and chirping birds connecting the inside and outside. The upper opening is covered with a large mirror, in which the flow of the Kiyotsu River in the distance is reflected. Visitors can feel the blessings of the earth on their feet, see the river flowing in the sky, and feel the journey that is about to begin.

Upon entering the tunnel, a long passage emerges with a sudden turn to the left. The lighting on the walls was partially removed during the renovation to reduce illumination, creating a dim cave-like atmosphere.

いうことで、まずは1996年に作られた姿のままとなっている。

更に先を進むとすぐに辿り着く第2見晴所。ここにはトイレを設置したのだが、このトイレの上部を球形に仕上げ、鏡面仕上げのステンレスで覆うことで、周囲の風景を映しこみ、まるでカメレオンのように空間の中に溶け込んでいる。内部は開口部に向けた方が誰でも使用でき、後部が女性専用となっており、前方のトイレは外部からは中が見えないが、内部からは外が透けて見える仕様になっており、トイレの中から見晴所に漂う人々の姿を見ながら、開口部の外に見える風景を眺めるという、特別な体験ができるようになっている。

そして第3見晴所。ここは最後のパノラマステーションの前の空間ということで、人々が漂う休憩スペースとして想定された。既存の照明器具の配線を利用し、オレンジ色の間接照明と、水滴のような形状をした鏡を組み合わせたモジュールを回転させながら配置することで、開口部に近い部分は外の風景を反射させながら徐々に内部に浸透させ、内部に行けば行くほど、徐々にオレンジの光が強く反射して空間を覆いだす。見晴所の中を歩くと、トンネル形状にそってさまざまな角度を持つ鏡に自らや他の来坑者の姿が何重にも反射する。

トンネルの最後は、青の光のトンネルが徐々に左に折れ、その先の第4見晴所（パノラマステーション）から入り込む光に誘われるようにして到着する。四つある見晴所の中で、他の3つの見晴所は峡谷に対し横から壁面を見るのに対し、唯一この「パノラマステーション」は峡谷に正対して見ることができる。見晴所の先端から目に飛び込んでくるのは、逆三角形に切り取られた空と、その下の清津川の流れという清津峡の静と動を感じることのできる風景。

見晴所自体に奥行きがあり、外の切り立った崖の遥か下に清津川が流れるために、見晴所に入った際には清津川は実際に見ることなく、激しい川の流れの音を聞きながら空間を進む。見晴所の床を水深の浅い水盤として、半円状の開口部から見える外部の風

The lighting is covered with a unique colored film that divides the tunnel into five different color zones, forming a series of vibrant tunnel sections. Each color is associated with a different vantage point or entrance. Therefore, rather than physically separating the tunnel from the vantage points, the entirety is connected as one seamless but gently sub-defined space.

From the center of the tunnel, an unrecognizable sound, not a song or a tune or even a language, wafts through the tunnel. The mysterious humming voice invites visitors deeper into the tunnel, as if they are leaving their present world behind.

The first platform retains its original appearance, allowing visitors to learn about the history of the Kiyotsu Tunnel. From this spot, visitors can observe the rough surface of the cliff and the rushing river. At the beginning, we planned on installing a new toilet at the first platform. However, after much discussion with local residents, and considering the tunnel's history upon which the renovation was based, we concluded it would be better for the visitors to walk through the tunnel with a natural flow. Therefore, we chose to preserve the first space as it was built in 1996.

Moving forward, visitors soon reach the second platform. The top form of the bathroom is spherical and covered with mirror-finished stainless steel, reflecting the surrounding scenery, and blending into the space like a chameleon. The side closer to the valley is unisex, while the other side is for women only. While the toilet interior cannot be seen from the outside, those inside the toilet can see through the walls to the surrounding tunnel. A special, surreal experience is created by witnessing both the scenery and the sightseeing people drifting around the tunnel space.

The third platform forms a resting place for people to drift around, leading towards the final viewing platform. By using the existing lighting system, and combining the dim tangerine lighting with drop-shaped

景を反射させ、円形状の風景となるようにした。逆三角形に見えていた空は、水面に映りこむことで峡谷の間を流れる川のように見えてくる。水盤には靴を脱いで中に入れるようになっており、人が居ない時や動いてない時には穏やかな水面に鮮明な像を映し出す。逆に人が多く、水面に波紋が立つような時には、映し出された像もまた揺らぐことになり、この場で来坑者がどのような行動をするかによって、この場の風景もまた変化する。

トンネルの壁面も、鈍い反射のステンレスパネルで覆うことにより、峡谷の切り立った崖面とその下に流れる清津川の風景をゆがめながら壁面と天井に鈍く映すことになる。自分の立つ場所によって、天井にうっすらと流れる川の姿を確認することができる。トンネル端部まで張られたステンレスパネルによって、外部と内部の境界線が曖昧になり、外部が内部に浸透するような効果を醸し出す。

空間の中には、目に見えない川の音が満ち、自らが立つ場所によって外部の空と川がさまざまな形で空間に反射されつつ、動くことでその像もまた変化していく。水の中を進むと、徐々に大きくなる川の音と共に、渓谷の底でダイナミックに流れる清津川の姿を目にとらえることになる。開口部に近づくにつれて、トンネルへと吹き込む風を身体に感じ、季節の変化とともに日々表情を変える清津峡の自然の風景を捉え、来坑者はそれぞれの心に、その時感じた感情と共にこの風景を焼き付けていくことになるだろう。

水盤の水は、清津川の水を使用しているため、夏場でも水温が低く、春や秋では長く足を付けているのが辛くなる程で、豪雪地であり、雪解け水の恵みが作り出す清津峡の風景を足の裏という自らの身体を通し感じてもらえるようになっている。

光や風、温泉水や沢の水、反射や色と、極めて基本的な要素のみを使用しつつ、清津峡やそれぞれの見晴所の特性に寄り添いながら、全体としてひとつの「旅」を紡ぎだすような作業であったと思う。芸術祭オープン後、「パノラマステーション」からそれぞれがどんな想いを感じたかなどを語り合いながら、色

mirrors, the area near the opening reflects the outside scenery and gradually penetrates the interior. The further one gradually travels inside, the stronger the tangerine light covers the space with its strong reflection. As visitors walk through the platform, they see multiple reflections of themselves and others in the mirrors at various angles along the tunnel.

Towards the end of the tunnel, the blue-lit space gradually turns to the left. Here, we see a "light at the end of the tunnel," with natural light in the distance enticing visitors towards the fourth and final platform. Of all the platforms, the fourth is the only spot offering a direct view of the canyon, with the other three only offering a view from one side. From the top of the viewing platform, visitors witness the sky cut out in an inverted triangle, as well as the flow of the Kiyotsu River, offering a sense of the stillness and movement in Kiyotsu Gorge.

The Kiyotsu River flows far below the platform and steep cliff. The platform itself is deep in width, meaning as visitors walk onto the platform, they do not see the river, but instead hear the rushing water beneath them. Inside the tunnel, the ground holds a shallow pond that reflects the external landscape seen through the semicircular opening, forming a circular landscape. The sky, which once resembled an inverted triangle, now resembles a river flowing through a canyon through its reflection in the water. Here, visitors can take off their shoes and step into the pond. When the pond is still, it reflects a clear image, while when the water ripples with the movement of people, the projected image also wavers. The scenery therefore changes depending on the movement of visitors.

The walls of the tunnel are also covered with stainless steel panels with a dim reflection, meaning the scenery of the canyon's steep cliffs and Kiyotsu River are distortedly reflected on the walls and ceiling. Depending on where the visitor stands, they can see the river flowing faintly on the ceiling. The stainless steel

のトンネルを通りそれぞれの日常に戻っていく人々の姿を見て、私たちの意図が良い形で人々に受け入れられているのだと実感することができた。

新緑に覆われ始める春の風景から、万緑の夏、紅葉が連なる秋、そして真っ白に染まる冬と、一年を通して日々多様な表情を見せ変化していく清津峡の風景。そこにそっと潜望鏡を伸ばして、外部を取り込み、それぞれの見晴所にて違った形で拡張することで、外の自然と訪れる人々の感情の間を反響しつつ、清津峡の魅力を巡る旅として、この土地に少しだけ新しい風景を作り出すことができたのではと思う。

panels stretch up to the end of the tunnel to blur the boundary between the outside and inside, allowing the outside to seemingly permeate the tunnel's interior.

The space is filled with the sound of the invisible river. The sky and river are reflected in the space in various ways depending on where the visitor stands, with the view changing as they move. As they walk through the water, the sound of the river gradually becomes louder, until they see the Kiyotsu River flowing dynamically at the bottom of the valley. As visitors approach the opening, they feel the wind blowing into the tunnel, capturing the natural scenery of Kiyotsu Gorge, whose expression changes every day with the seasons.

The water in the Light Cave is drawn from the Kiyotsu River, meaning the temperature is low even in summer. In the spring and autumn, it is a challenge for visitors to keep their feet in the water for too long. The scenery of the Kiyotsu Gorge, created by the blessings of melting snow, is felt through the visitor's own body, and the soles of their feet.

Now completed, the project continues to change over the course of the year along with Kiyotsu Gorge itself. The tunnel travels through the cycle of a verdant spring, an emerald summer, an auburn autumn, and a white winter. The periscope, a channel through which sensibilities and nature communicate, serves to bring into focus the landscapes visible from each platform. The outside is invited into the tunnel, intensifying the unique experience provided by the gorge. In this sense, it could be said that we have created entirely new sceneries in this landscape.

エントランス施設　The Entrance Facility

エントランス施設の足湯の屋根の素材は初期の提案からかなり変化しています。最初は板張りの案でしたが、MADとグリーンシグマは耐久性や降雪などさまざまな要素を配慮し、素材についていろいろ研究してくれました。

予算的な制限がありましたので、最終的には一般的な屋根の素材になりました。普通は、屋根は平面的に葺きますので、当然そうなるかと思っていましたが、彼らは少し段差をつけたデザインを考えてくれました。こういうところが、やはりMADはすごいと思います。限られた条件のなかでできるだけ質の高いデザインを見せる工夫をし、細かいところもおろそかにせず徹底してデザインを練ってくれました。

完成してみると、とてもモダンでありながら、景観の中に溶け込んでいるようです

——山本勝利（十日町市中里支所地域振興課）

Initially, we intended the roof of the entrance facility to be made of wood. Considering the local climate, and in an effort to improve durability, MAD and Green Sigma invested substantial time discussing the choice of material. The final result ultimately changed significantly from the initial plan.

While we ultimately chose a common roofing material due to engineering and budget constraints, the construction method exceeded my expectations. Rather than installing the panels in a traditional way, the builders made use of a slope difference in the design. Reflecting back on the project, I admire how MAD can deliver quality designs within such constraints. Their commitment extends to the tiniest detail, with the design carefully considered throughout.

Once completed, the design evoked a contemporary feel, while still in harmony with the surrounding natural environment.

— Yamamoto Shori (Section Chief, Nakasato Branch of Tokamachi City)

40-41. 撮影：岡本裕志　　　　　　　　　　　40-41. Photography by Hiroshi Okamoto

最初にエントランス施設の足湯のデザインをMADから受け取った時には、鏡の反射はどこの景色を狙ってるのかよく分かりませんでした。自分としては清津峡の渓谷を狙っているのかと思いましたが、実際にデザインを進めていくなかで、建物にかなり近い空間を映すという意図が見えてきました。それからは現場の各担当のみんなと懸命に努力して実現させていきました。

鏡の角度を調整する時には、早野さんとマ・ヤンソンさんに実際に現場まで来ていただき、どこを、どのように映すのか確認していただきました。

エントランス施設の屋根はシンプルなデザインのように見えますが、実は台形の屋根を二つ重ねて構成しており、角の部材の寸法が全て違います。

5メートルの積雪に耐えうる屋根の設計にしているため、屋根に強度が要ります。そのためにパイプに多くの細かい材料を溶接する必要があり、熱のために材料が変形してしまいました。鉄工場でそれを調整して現場に運び、ようやく組み立てることができました。このように苦労をしたところもあります。

——瀬戸智（グリーンシグマ取締役、技術総監）

When I first saw the design for the foot spa at the entrance, I did not understand what specific scenery would be reflected. At first, I thought it would be elements near the stream. As the idea evolved in detail, I understood that the reflection was of the landscape closer to the facility, and I worked with representatives from the different teams on-site to realize the idea.

To make the effect align as close as possible to the original intent, Mr. Hayano and Mr. Ma adjusted each mirror themselves. The specific part of the surrounding being reflected, and how it is reflected, were created by them personally.

The two sides of the roof may appear as two simple trapezoids joined together. But, in effect, every piece of the material has a different length and shape.

The design of the roof must be capable of withstanding 5 meters of snow. To ensure the roof is sturdy enough, we fused many small elements together inside the pipes. However, the process produced heat that caused the material to deform. The elements therefore had to be readjusted at the factory which produced the steel structure, and assembled later on-site. A great amount of thought was invested in issues such as these.

— Satoshi Seto (Technical Director, Board Member of Green sigma Co.,Ltd.)

ひとつひとつの角度、1枚1枚の鏡のすべては私たちと施工スタッフが現場で少しづつ決めていったものです。施工現場の条件に限りがあり、設計時の位置と角度は施工時から変化しています。それぞれのポイントの視角から鏡のなかに見ることのできる景色を精密に測るため、こうした微細な変化を事前に制作した模型に反映させる必要がありました。

——早野洋介

The angle of each mirror was adjusted little by little on-site with the help of our construction team. Due to the limited construction area, the position and orientation of the building differed slightly from the plan. Therefore, the angles of the mirrors also needed to be readjusted, as the actual reflections seen from different positions were an important consideration when we initially designed the building.

— Yosuke Hayano

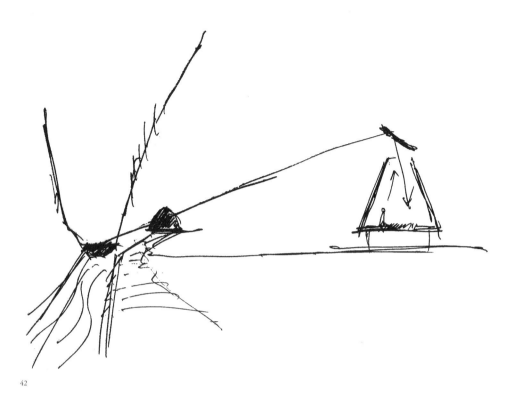

42

42. 川はトンネル入口付近を流れているが、鏡がその風景をとらえ、足湯空間にいる人の目に映し出す

42. The river flows by the entrance of the tunnel. This view is captured by the mirror, and witnessed by visitors sitting in the foot-bathing area.

85

43

43. マ・ヤンソンと早野洋介が施工現場にて鏡の角度を調節
©MAD Architects

44. エントランス施設内部パース

45. 鏡の位置と反射角度の関係から一階からの階段の位置と空間内の動線
が決定され、それに伴い一階での階段の開始位置と、受付スペースの形状
も決定される

43. Ma Yansong and Yosuke Hayano adjusting the mirror angle at the
construction site. ©MAD Architects

44. The entrance facility rendering ©MAD Architects

45. In order to make the entire visiting experience more natural, the location of
the staircase is determined by the position of the mirror and the direction of the
reflection. This also determines the starting point of the ground floor staircase
and the arrangement of the reception space.

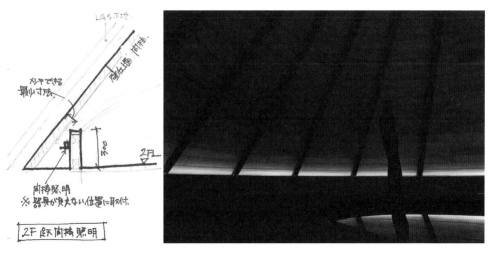

LGS下地

メジでとる
最小寸法

壁仕上面 間柱

2F床
▽

間接照明
※器具が見えない位置に取付.

2F 欄間接照明

44

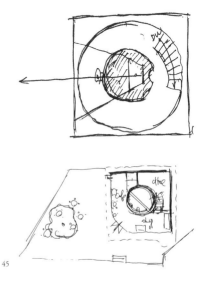

45

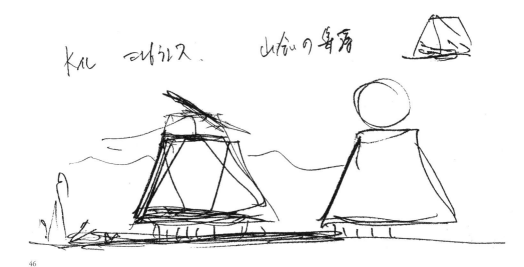

46

46. エントランス施設は敷地にかかる法規制の為、伝統的な切妻屋根を採用する必要があった。私たちはこの条件を利用し、屋根の開口部に鏡を設置することで、ユニークな空間を作り出した

46. The entrance facility needed to feature a traditional triangular roof due to local regulations. MAD explored how to modernize the form by adding a floating mirror to the roof.

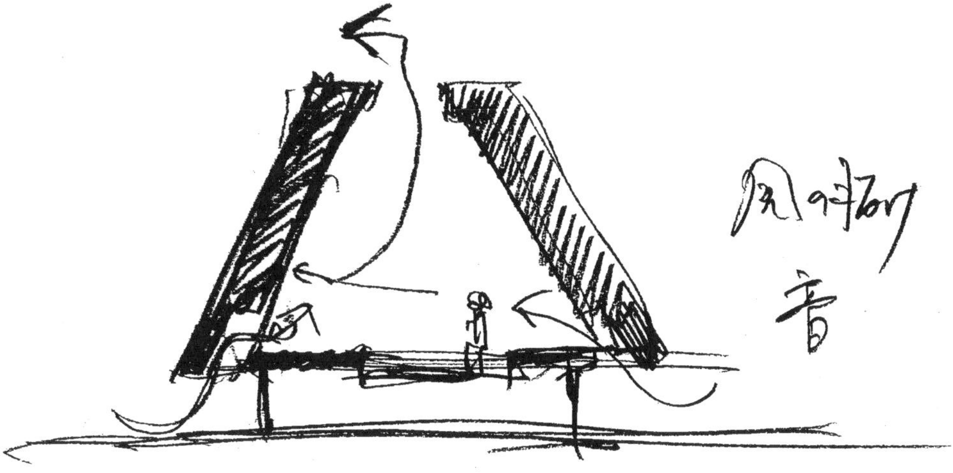

47

47. 一階の周辺空間は大屋根に覆われた縁側空間として機能する。二階は足湯の温泉水供給設備を鏡と同じサイドに配置することで、人々が鏡に対面する方向に座るよう自然と誘導し、そこから鏡に反射する外部の清津川の風景を眺めることになる

47. The surrounding area on the ground floor becomes a gray space under a large roof. On the second floor, by positioning the footbath water supply beside the mirror, people are naturally led to sit opposite it; facing the mirror to observe the view of the river in its reflection.

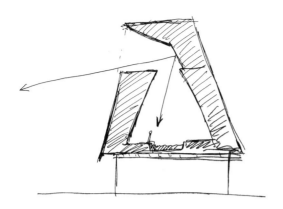

48

48. 現場にて鏡の角度を調整 ©MAD Architects

48. Adjusting the mirror angle on site ©MAD Architects

トンネル

The Tunnel

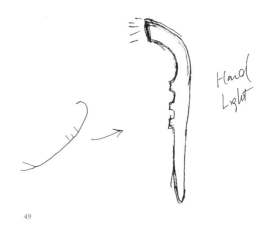

49

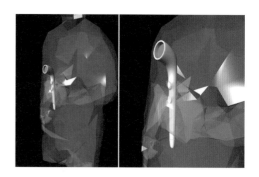

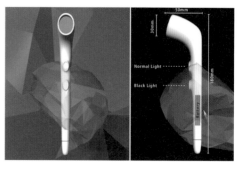

49. 懐中電灯の形状はトンネルの形状からきていて、側面に3つの押ボタンがあり、先端は電灯になっている ©MAD Architects

49. The shape of the portable lamp is inspired by the abstract form of the tunnel, with three buttons on the side and a lamp head at the top. ©MAD Architects

かつての清津峡トンネルは、訪れた人は最初から終わりまでまったく変化のない照明のもと、黙々と700メートルほどを歩き、ようやく見晴所に至るようになっていました。私たちは壁面の照明を減らしたうえで、さまざまな色の特製フィルムをつけ、トンネルを5つの異なる色の区域に分け、色が織りなすグラデーションのトンネルへと変化させました。

初期には既設の照明をすべて取り外し、私たちがデザインした懐中電灯を訪問者に渡し、人々はそのトンネル形の灯を持ち、本人と同伴者が手元の灯をあたかも潜望鏡のようにして進んでいき、人の全感覚が最大限に鋭敏になるような体験を計画していました。けれど最終的には安全を考慮してこのアイデアはあきらめたのです。

———早野洋介

Before the renovation, visitors would walk more than 700 meters in a continuously lit tunnel to visit the different platforms. We detached some of the lights, decreased the luminance, and covered the light tubes with specially made tinted films. Now, the colors divide the tunnel into five sections, and together they transform the space into a spectrum of light.

Initially, we wanted to remove all lights inside the tunnel. The vision was to give each visitor a handheld light in the same shape of the tunnel, reminiscent of a periscope. Visitors would depend only on their lights, and those held by their friends and other visitors, to see the road. Senses would become heightened and vigilant, resulting in an intensified experience. Ultimately, we rejected this idea for safety reasons.

— Yosuke Hayano

見えない泡

当初は第1見晴所に公共トイレを新設する予定でしたが、プロジェクトの過程で地元住民のみなさんとの対話を重ねるなか、初めてこの場所を訪れる人にも、これまでの清津峡トンネルがどんな経緯で設置され、今回のリノベーションを経てどう変化したかを理解してもらえるよう、この第1見晴所は何も手を加えず、設置された1996年のままとする決定をしました。

——早野洋介

The Bubble

The initial plan was to have the restroom on the first platform. However, after learning more from the locals, we altered the plan. We wanted visitors to see that the project was developed based on the history of the gorge. Hence the first platform now preserves the tunnel's initial condition, which was first built in 1996, allowing visitors to look back upon the history of the gorge.

— Yosuke Hayano

MADの提案通りのアクリル素材を使ったドームのトイレを作る会社は日本で二つしかありません。そのうちひとつに委託を受けていただきました。

最初、メーカーの担当者の方は今の形よりさらに膨らむ、もっと天井の高いものができるだろうと言い、できるだけMADのデザインに近いものを実現しようとしていました。でも、実際にドームの制作を始めると、膨らみがうまくいかず、膨らませる途中で、横に亀裂が入り、ドームが萎えてしまいます。そんな状態が何回も続き、計7回チャレンジしましたが、6回失敗しました。本作が唯一、成功したものなのです。

ドームの表面には「銀鏡塗装」という特殊な塗装を使い、ハーフミラーにしました。実際、そこに入ってもらうと、中からは清津峡の景色が透けて見え、外側からは中が見えない状態を最終的に実現させています。

——瀬戸智

There are only two companies in Japan that can produce the acrylic globe required by MAD. Only one of those accepted our commission.

The original version of the Bubble was of a more inflated shape, with a higher topped dome that aligned closer with our initial vision. However, once production began, we found it frustratingly difficult to make the dome. We tried seven times in total. For the first six trials, the dome collapsed halfway through the inflation process. What we see in the tunnel now is the successful seventh version.

For the surface of the dome, we used a special coating process called "spray-on chrome" to create a see-through mirror effect. As a result, visitors inside the restroom can still see the natural scenery outside, while being hidden from the view of the visitors on the platform.

— Satoshi Seto

50. ©MAD Architects

51

52

トンネルの天井高はとても低く、また見晴所での作業可能なスペースも非常に限られていました。何度も検討を重ねた結果、唯一可能な方法は工場で製作をして運び込み、組み立てだけを現場で行う方法でした。トイレを覆うアクリル製のシェルは特殊な塗装に覆われているため、万が一作業途中にトンネルの天井に触れて塗装に傷がついたりしたら、再度作り直すことになります。その時間は残されていなかったため、現場への運搬から設置、完成まではとても緊張しました。

──宮本一志（元MADアーキテクツ所員、
プロジェクト・アーキテクト）

The ceiling height of the tunnel is exceptionally low, leaving the construction team with limited space to work within. Transporting the only successful dome we produced onto the platform was more than thrilling. The ceiling could have easily scratched the coating; and we hardly had the time and money to make a new one.

— Isshi Miyamoto, Former Architect at MAD Architects, Project Architect

53

54

水盤の鏡

水盤についてひとつ面白い話があります。本当に鏡の状態になるのかを心配した市役所の担当の方が、子ども用の小さいビニールプールに黒いビニールのシートを貼り、水をはって現場に置いて実験し、本当に映るかを確認しました。

面白いですよね。私は「底を黒くすれば必ず映りますよ」と言っていましたが、言葉ではなかなか納得できないもので、市役所の方が自ら実験し逆さに映ることを確認したのです。

——瀬戸智

The Light Cave

There was an interesting episode during the construction of the light cave. The city hall staff who oversaw the project were concerned that the reflective effect of the pond may not turn out as planned. To test the effect, they covered the bottom of a children's inflatable pond with black plastic bags and filled the pond with water. They then brought the pond to the site to verify the reflection.

I had reassured them that as soon as the bottom became black, a reflection would naturally form in the water. Nonetheless, they still insisted on an actual demonstration. I found it quite interesting that, for them, verbal assurances were not enough, and they carried out an experiment to verify the effect.

— Satoshi Seto

57

清津峡のあらまし

1941-1996年（昭和16年-平成8年）　国立公園に指定された後は、高度経済成長期も重なり、毎年、秋には多くの観光客が訪れるようになった。

2016-2018年（平成28年-平成30年）　2016年、リニューアル工事を開始、清津峡トンネルの年間来場者数はおよそ延べ5万9千人、全盛期の三分の一にまで減少。

2018年（平成30年）　統計によれば、リニューアルオープン後の2018年（平成30年）、清津峡トンネルの年間来場者数は延べ18万3千人に達し、リニューアル前の三倍となった。2019年（平成31年）、年間来場者数は延べ29万6千人となった。

リニューアルの大成功により、大地の芸術祭の開催年ではない2019年のゴールデンウィーク、お盆にも多くの観光客が訪れ、過去の最盛期にも例を見ないほどの大盛況となった。

同年「Tunnel of Light」は、文化庁および日本芸術文化振興会により、古代文物から現代文化アートまでに至るなかから日本の文化を代表する5作品の一つに選ばれ、グローバルオーディエンスむけのPR動画に選ばれ登場した。

『ジャパンタイムズ』、『Nikkei x TECH』、『Designboom』、『Dezeen』、『周末画報』、『三聯生活周刊』、『人民網』など国内外のさまざまなメディアで取り上げられ、中国の建築家マ・ヤンソン率いるMADアーキテクツによる、日本における地方文化と経済を振興するアート作品として報道された。

突如発生したCOVID-19は多くの人々の生活に影響を与え、アートと人類と自然のつながりの重要性がますます浮き彫りになった。当初2021年開催予定だった第8回大地の芸術祭越後妻有アートトリエンナーレは延期となった。北川氏の招待を受け、MADは「見えない泡」のある空間に改造を加え、新作品「Flow」を設計、引き続き自然をテーマにした創作を行うと同時に、新たに生まれ変わる希望を人々にもたらした。

次なる芸術祭は、徹底した感染対策のもと、「地域おこし」につながる取り組みをさらに発展させ、作品展示のインターネット配信など、現地に来訪できない人たちも芸術祭を楽しめるような取り組みを行っていく予定である。

About Kiyotsu Gorge

1941-1996 (Showa 16 to Heisei 8) As Kiyotsu Gorge became a national park and a natural heritage site, and with Japan experiencing an unprecedented economic boom, the gorge became an autumn destination beloved by tourists.

2016-2018 (Heisei 28 to Heisei 30) In 2016, prior to the renovation project, the annual number of visitors had dropped to around 59,000, about a third of the number in its heyday.

2018 (Heisei 30) According to statistics, the number of visitors in 2018 reached 183,000, thrice as before the renovation

Thanks to the success of the renovation project, Kiyotsu Gorge welcomed a large number of visitors during the Golden Week and Bon Festival in 2019, unprecedented for a year without the art Triennale.

In 2019, the Kiyotsu Gorge Tunnel appeared in a global presentation video made by the Japan Cultural Expo. In the video, the reinvented tunnel represented the culture of Japan alongside four other art forms, featuring national treasures and traditional arts that continue into the modern day. `

The Japan Times, Nikkei x Tech, Designboom, Dezeen, Modern Weekly, Sanlian Life Weekly, People's Daily Online, and other media covered the art project led by Chinese architect Ma Yansong and its contribution to revitalizing the rural area.

In 2019, the annual number of visitors to the Kiyotsu Gorge Tunnel reached 296,000.

2021 (Reiwa 3) The power of the project lies in its relationship between art, humanity, and nature; a spirit which became even more relevant during COVID-19. The wider 8th Echigo-Tsumari Art Triennale, originally scheduled for 2021, was postponed due to the pandemic. However, by invitation of Mr. Kitagawa, MAD reinvented the space around the Bubble with their new exhibition "FLOW." Continuing along the wider project's theme of art, humanity, and nature, the exhibition offered a new hope for a normal life beyond the pandemic.

58

59

60

61

62

64

65

66

67

65. リニューアル前の第一見晴所 ©MAD Architects

66. リニューアル前の第二見晴所、リニューアル後「見えない泡」
©MAD Architects

67. リニューアル前の第三見晴所、リニューアル後「しずく」
©MAD Architects

65. The first viewing platform before the renovation ©MAD Architects

66. The second viewing platform before the renovation, later transformed as the "Bubble" ©MAD Architects

67. The third viewing platform before the renovation, later transformed as "Drop" ©MAD Architects

68

68. リニューアル前の第四見晴所（パノラマステーション）、リニューアル後「ライトケーブ（光の洞窟）」©MAD Architects

68. The fourth viewing platform before the renovation, later transformed as the "Light Cave" ©MAD Architects

69

70

69-73. ©MAD Architects

71

74

72

75

73

76

74.「Tunnel of Light」は、2019 年文化庁による「日本博」において、古代から現代における 日本の伝統と芸術を代表する 5 作品のひとつとして選ばれ、日本文化の代名詞となった

75. 日本の建築雑誌『日経クロステック』掲載

76. 韓国の雑誌『Heritage Muine』掲載

74. In the 2019 "Japan Expo" global campaign video, the "Tunnel of Light" was featured with four other traditional arts that had been passed down from ancient times to the present, to represent Japanese culture.

75. Japanese media coverage from *Nikkei x TECH*

76. Coverage in the Korean magazine *Heritage Muine*

「Tunnel of Light」による
地域活性化

Local Revitalization Brought by the Tunnel of Light

77

関口芳史 十日町市長

Told by Yoshifumi Sekiguchi, Mayor of Tokamachi City.

Since 2009, Mr. Yoshifumi Sekiguchi has served as the Mayor of Tokamachi. He is also the Chairman of the Executive Committee of the Echigo-Tsumari Art Triennale. The following interview was conducted in 2019.

関口芳史氏は2008年より十日町市長を勤め、「大地の芸術祭」実行委員会の委員長も兼務している。1959年に十日町に生まれ、15歳まで地元で育つ。高校、大学は東京に出て学び、東京の会社に就職。36歳でその会社を辞め、家業を手伝うために故郷に戻った。十日町でしばらく働いた後に、市役所で現在の副市長に相当する役職で勤務し、新潟県の三条市の収入役としても勤務していた。

Born in 1959, he lived in Tokamachi until the age of 15, at which point he moved to Tokyo to attend high school and university. After graduation, he worked in a company in Tokyo. At the age of 36, he left the company and returned to his hometown. He first worked in Tokamachi and then in the city hall. His title in the city government corresponded to today's deputy mayor. He has also worked as an accountant in public services in Sanjo City, Niigata.

清津峡のトンネルには改修する前にも何度か行っていましたが、その時には最後の見晴台までの距離をかなり長く感じました。そして、時の流れとともに、清津峡のトンネルもひどく古びていきます。これを一新させよう、芸術の力で清津峡のトンネルを生き返らせたいと強く願うようになりました。MADと提案を検討するなかで、さまざまなアイデアが生まれ、何度も討論を繰り返して最後に今の方法になりました。図面で見た時に想像したよりもはるかに素晴らしい結果になったと思っています。

I visited the tunnel before it was renovated. Walking through the entire tunnel felt long. Over time, the tunnel began to deteriorate. I had a strong urge to bring new life to the tunnel through art. Later, when discussing the project with MAD, we came up with all kinds of ideas. After a series of discussions, we agreed on a final design. The actual result is far more stunning than what the models and blueprints showed. It exceeded my imagination.

改修後はトンネルに出かけるのは、ワクワクと興奮させられるものとなり、それぞれの見晴所の風景を満喫し、知らず知らずのうち最後の素晴らしい水盤鏡に到着するようになりました。水から反射される景色、ステンレスの壁が反射する風景は想像を超えたもので、特に逆光になり、そこに人がいてシルエットとなる効果は本当に素晴らしいものです。

Walking through the new tunnel is at once exhilarating and relaxing. As one lingers over the different sceneries from one platform to the next, they soon arrive at the final, and also the most impressive design: the light cave. The view formed by reflections in the water and on the stainless-steel wall surprised me. Because the space is backlit, we see one another through silhouettes.

水盤には美しい自然が映り込んでいます。越後妻有の地域の特徴は四季がはっきりしていることですが、春や夏は美しい緑、秋は紅葉があり、冬は白い雪です。地域

At the light cave, the breath-taking beauty of nature floods inwards. The climate at the Echigo-Tsumari region has four seasons, which is aptly expressed in the work of MAD. Reflections of nature and silhouettes

の特徴を際立たせてくれる作品だと思います。そこに周りの風景と楽しんでいる人々のシルエットが映り込んでいます。人の顔ははっきり見えなくても、体が喜びを表現している感じにはなっています。訪れる人が越後妻有の美しい自然を心から満喫しています。この効果は、さすがMADだと思います。お客様をご案内することも多いですが、私自身も楽しんでいて、行くたびに本当に感動しています。トンネルの750メートルの距離があまり感じられなくなりました。

大地の芸術祭でさまざまな作品が生まれましたが、越後妻有において「Tunnel of Light」は最も人気のある作品です。お客さんは多く、インスタグラムを中心にネットでも非常な影響力があります。これまでの清津峡のお客様は紅葉を楽しむ人たちで高齢の方が中心だったかと思いますが、今は若者や女性などまったく違うお客様にアピールできていますね。

大地の芸術祭のさまざまな作品の中でも、「Tunnel of Light」は最も注目されている作品だと思います。テレビ、新聞、雑誌など一流メディアからの取材が多いです。最新技術で映像を撮ってくれたあと、少し光の入り方が違うからと再撮することもあり、そんなことから皆さんがとても注目してくれているのがよくわかります。

メディアはまず作品の芸術性に注目しています。「Tunnel of Light」には斬新なビジュアル効果があり、今までにない体験ができることで注目されていると思います。また、ストーリーを探求する人たちはいろいろな過去の歴史を調べてきます。「不幸な落石事故があって、清津峡を十分に楽しめなくなった。そしてやむを得ずトンネルを掘った。はじめはお客が多かったが、次第に減ったお客様。MADのおかげで、再び大ブレークした」というストーリーに感動している方も多いです。

清津峡は十日町の玄関口です。そうした意味で「Tunnel of Light」は、芸術作品として力があります。そのおかげで十日町、越後妻有の集客力が非常にアップし、より多くの観光客がこの地を訪れ、地域の活性化に貢献しています。

2018年に3年に1度の「大地の芸術祭」が開催されました。

of people enjoying themselves both activate and complement one another. Although people's facial expressions are dimmed, the cheerful and joyous mood is conveyed by their gestures and movements as they heartily enjoy the wonderful landscape of Echigo-Tsumari. MAD's design deeply moves me. Nowadays I often bring guests to the tunnel and have enjoyed the trip myself again and again. The 750 meters has never seemed so short.

On this land, the Echigo-Tsumari Art Field has delivered many artworks. Among them, the Tunnel of Light is undoubtedly the most celebrated in the Echigo-Tsumari region. It has been popular among tourists as well as on Instagram and other social media. Before, the visitors were mainly elderly, here to enjoy viewing the autumnal maples. Now the gorge has attracted younger people, as well as more female visitors.

Among the many works at the Triennale, the Tunnel of Light stood out and drew attention from top TV channels, newspapers, and magazines. They came with the latest technology to film the tunnel. Sometimes they would film a second time simply because light entered the tunnel differently. It also made me realize how much attention the project was getting from wider society.

The primary focus of the media was the project's artistic aspect. The Tunnel of Light has created unparalleled visual effects. Another focus has been the history; a rockfall accident occurred, leading to the total shutdown of the gorge. The tunnel was a remedy. Tourists came back for a brief period, but the curiosity soon died down. The number of visitors continued to decline. Now, thanks to the work by MAD, the number of tourists has again increased to an unprecedented level. Many people are touched by this story.

For Tokamachi, Kiyotsu Gorge is like its vestibule. Attractive as it is, the Tunnel of Light has significantly boosted the popularity of the Tokamachi City and

当時たくさんのお客様が訪れました。2019年には芸術祭は開催していないのですが、2019年に訪れたお客様のほうがさらに多いことは、この作品の芸術性の証明だと思います。

清津峡は国立公園の中にある本当に美しい峡谷です。そこには奇岩がたくさんあって、面白い形をしています。自然と町も本当に美しいところですが、残念ながら落石事故があり、谷底に下りていくことはできなくなってしまった。せっかくの財産にアクセスできなくなり、その時には皆さん本当に悲しんだと思います。

清津峡トンネルの建設を決断した旧中里村の山本村長は、私の友人です。あんなに素晴らしい資源を十分生かせていないことを、彼もとても残念に感じていました。「トンネルを掘ったけど、トンネルからでは、本当に見たいものが見られないのではないか、なんとかもう一度、下りられるようにしてほしい」と言っていました。「危険だろうけどね。危険をうまく回避して、訪れる人が谷底に行けるようにしてほしい」と言っていました。

そして、彼は亡くなりました。彼の言葉が遺言のようなものになりまして、それは私の心の中にずっと残っています。今回、トンネルは素晴らしい芸術作品として生まれ変わりました。山本さんの思った通りにはならなかったかもしれませんが、たくさんの来訪者の方に清津峡を楽しんでいただくことは実現できました。その意味では山本さんの思いに応えることができたと思い、嬉しいです。本当によかったと思います。

the Echigo-Tsumari region. As more and more visitors come, the local culture and economy is also further revitalized.

The latest iteration of the Triennale took place in 2018. However, the number of visitors continued to increase in 2019. This proves how influential the artistic aspect of the work is.

Kiyotsu Gorge lies inside a national park. The unique geological landscape, and its formation, intrigued people. Alongside the reverent nature, rural life went on for centuries. However, after the rockfall accident, the gorge had to close. People could no longer enter the gorge and were saddened by the loss.

The former leader of Nakasato Village was a friend of mine, who set out to build the tunnel of Kiyotsu Gorge. For him, the loss was especially heavy; that nobody had access to the beautiful scenery anymore. Although a tunnel was later built, the splendor was greatly reduced, with people only witnessing the gorge's magic from a distance, or inside the tunnel. He genuinely wished that one day the gorge would reopen to visitors. "Though dangerous, we still need to find solutions regarding the dangers posed by the gorge so that people can cross the gorge again," he said.

After he passed away, his dying wish moved closer to my heart. Through the outstanding works of art, the tunnel has obtained a new life. Although it differs from what Mr. Yamamoto would have wanted, more and more people now get to share the extraordinary experience. In a way, Mr. Yamamoto's wish has been fulfilled. Knowing this comforts me.

「ぜひここに見にきてください。」

78

桑原清 清津峡温泉組合長、清津館オーナー

桑原清氏は清津峡温泉組合長を勤め、温泉旅館も経営している。温泉組合の組合長として、重大な責任を担い、地域の自然、住民の生活、温泉施設などを守っていくため、行政の人々と調整をしながら、地元の人たちと協力してさまざまな活動をしている。温泉旅館も経営しているため、その日々はとても忙しい。旅館は一日中営業しているが、その間に組合や観光協会の仕事をしたり、多方面に関わっている。清津館はただの旅館ではなく、地域の拠点のような存在だ。地元の人たちは何かあると、とりあえずここに相談にくる。客好きな桑原氏は、ぜひ清津峡を訪れて、自然を満喫し、温泉を体験し、米、地酒、山菜を食べてみてほしいと語る。

今から200年ぐらい前に、清津峡渓谷に温泉が発見されました。そこからここまで温泉を引いてきました。第一号の旅館はうちの隣で、それからいまの近所の人たちが引っ越してきて、現在の温泉街が形成されました。

当時は、登山や温泉を目的とした観光客がかなり賑やかに訪れていました。これは遊歩道が閉鎖する以前の、昭和時代までの歴史です。当時は川側に約12キロの縦走の登山道がありました。特に10月の紅葉期には、日本のあちこち、そして海外をも旅する旅行好きの方もここに来られていました。

以前の道路では、川側の非常に狭いところを歩いたため、柱状節理の岩場が迫ってくるような、とても迫力のある景色が見られたのです。これが元来の清津峡の姿なんですね。

でも事故で登山道は通行できなくなってしまった。

"Please stop by"

Told by Kiyoshi Kuwabara, Head of the Kiyotsu Gorge Onsen Association, Owner of the Kiyotsu-Kan Onsen Hotel.

Mr. Kiyoshi Kuwabara has many responsibilities, as the head of the local Onsen association and the owner of the Onsen Hotel. From early morning, he is busy running the hotel, and during small breaks takes care of matters in the Onsen Association and the Tourist Association. He rotates between different roles and duties, be it communicating with government staffs, or recruiting residents to help him protect nature, promote the well-being of local life, and maintain a good condition of the onsens and pipes.

Mr. Kiyoshi Kuwabara's hotel is not just a hotel for customers. Over the years, it has become a gathering place. Villagers stop by to speak to Mr. Kuwabara when they need help or advice. Proud and accommodating, Mr. Kuwabara and his hotel will always welcome friends and guests to Kiyotsu Gorge with onsen, regional rice, sake, and mountain greens.

About 200 years ago, hot springs emerged in the gorge, the water from which people channeled to the village. The family who first started the hotel business was our neighbor. Naturally, the hotel became the starting point of Onsen Street. As more and more people came and settled down, the street gradually became what we see today.

Before the hiking trail was closed in the Showa era, the village was busy with guests and visitors who were mostly hikers or onsen-goers. People could walk for 12 kilometers, along the river and through the forest. The autumnal maples were another attraction. In October, visitors from all over Japan, and travelers coming from abroad, visited Kiyotsu Gorge.

The original hiking trail was tightly set between the river and one side of the rocky cliffs. Walking through the gorge, people were in awe of the scenery. When entering the gorge was prohibited, such intimate contact with nature was no longer possible.

「では、どうしたらいいのか？」と皆さんと考え続けました。ここは国立公園の特別地域内で国の名勝・天然記念物指定の区域ですので、建築物は絶対に許可されません。皆さんと検討を重ねた、国にトンネルの建設について問い合わせました。その結果、「トンネルであれば、山体の内部だから」と計画は許可されました。

トンネルが完成した当時は、とても話題になり、観光客の方も大勢訪れました。トンネルは、景色を見るためのトンネルになり、目的は達成されました。でも私としては、以前の登山道からの景色を見ているので、トンネルからの限られたアングルになるのは、少し残念だとは思っています。でも、ともかくも、なんとか景色は見られるようになったのです

実はトンネルにもメリットもあります。もとはこの峡谷は、冬に雪が降ると入れなかったんですね。でもトンネルならば、冬でも中に入り景色が見られるという利点ができたということなんですね。

それでも岩場を見る観賞用のトンネルだったために、その内部は単調で変化が乏しくで、観光客の方は次第に減りつつありました。

今回、「大地の芸術祭」という絶好の機会に、アーティストの方に改修いただき、このトンネルの作品は景色がとても話題になりました。完成間近のプレオープンで、初めてトンネルの中に入り、一番奥の水盤鏡を見た時、自分の目が信じられないぐらいでした。このアイデアは本当に斬新です。

「ライトケーブ」は皆さんと同じく私も大好きなのですが、特に私がいいなと思うのは、オレンジ色の照明部分です。光と鏡による空間作りは想像を超え、ただ驚いています。

トンネルの改修後に、観光客の方がとても増えました。うちの旅館はトンネルと結びつけたプランも立ち上げ、最近は若い方たちがたくさん参加するようになっています。

After the accident, the hiking trail was closed for eight years. However, residents did not give up finding an alternative. As the area is a designated region for national attractions and natural monuments within the National Park Special Area., any construction over land would be denied. Ultimately, the residents applied to the state for permission to build a tunnel. Since the main structure would be inside the mountain, the proposal was granted.

When the tunnel was completed, news about the development was widespread for a time, which helped to promote the gorge and attract more visitors. By building the tunnel, it became possible for people to come closer to the unique landform of Kiyotsu Gorge. The drawback was that the tunnel confined both the view and the experience. Personally, it still left me with a feeling of loss, in comparison to the time of the hiking trail. Nevertheless, we gained access to the gorge again.

But there was one advantage to having a tunnel. Previously the gorge was forced to close in winter due to snow. With the tunnel, people can safely visit and experience the winter scenery of the gorge.

Nevertheless, the tunnel built for viewing the rocky cliff offered only a very basic interior. The atmosphere of the tunnel itself was monotonic, and initial popularity in the experience quickly dwindled away.

The Echigo-Tsumari Art Triennale presented a great opportunity. Artists were commissioned to renovate and reinvent the tunnel, and ultimately, the entire visiting experience. The artworks and the scenery became heated topics. At the pre-opening, we entered the new tunnel for the first time. When I arrived at the end of the passage, the light cave, I could not believe my eyes. Making a pond into a mirror was such a brilliant idea.

Apart from the much-applauded light cave, the

皆さんぜひここにお越しいただき、自然に触れて、トンネルの全体を満喫していただき、あるいは川に下りて、自然に触れていただきたいですね。エントランス施設の足湯では、くつろぎつつ、鏡に反射されている風景を見上げて楽しんでいただきたいです。ここの温泉の泉質は、肌にいい泉質なんですね。ぜひまた皆さんに体験していただきたいと思っております。

この辺りはお米、お酒、そして山菜がとても美味しいですよ。こうした食べ物は四季を感じさせるものですね。春の新緑の時期には、いろんな山菜が出て、自然食が食べられます。秋はコシヒカリの新米が食べられる時期です。新潟県は、相対的にお米が有名ですが、特にこの地域は山菜やお米などが有名です。皆さんぜひお越しになって味わっていただければと思っております。

tangerine section also struck a chord with me. The space was transformed by lights and mirrors; the change was unexpected.

After the renovation, the number of visitors increased. At the hotel, we also developed programs and events related to the tunnel. As a result, we are beginning to see more young people visiting.

I strongly encourage friends and families to come and visit. You can connect with nature, walk through the tunnel, and venture near the river. Once you are here, enjoy a foot spa at the entrance, muse over the reflections above the skylight, and experience the nourishment of the onsen water. Be sure to come and try for yourself.

The rice, sake, and freshly foraged mountain greens are seasonal, hence appetizing. As spring breathes life into the land, the mountains sprout all kinds of greens and natural ingredients. With autumn comes the newly harvested koshihikari rice. The entire Niigata prefecture, and here especially, is well-known for quality rice cultivation.

「光のトンネル」の物語　Stories of the Tunnel of Light

79

村山久江　村山荘オーナー

Hisae Murayama, Owner of the
Murayamasou Hotel

私の生まれは十日町です。嫁いで3日後に、突然50人も泊まりにきましたよ。10月の最高の紅葉時期で、登山道に沿ってグループの人たちがやってきて、「どこでもいいから泊めてください」という感じでした。嫁にきたばかりで、「こんなに毎日忙しいの？」と、ものすごくびっくりした思い出があります。

それからはだんだんとお客様が減ってきました。昭和63年（1988年）に渓谷の奥で思いがけない事故があって、渓谷の登山道が通れなくなりました。その前に、昭和59年（1984年）の災害でうちの民宿が倒壊して、昭和61年（1986年）に旅館を建て替えました。主人が食堂を経営したいということで食堂も始めました。

お客様が10時にチェックアウトすると、食堂の準備をします。食堂は11時に営業を始めます。ものすごい忙しさでずっとやってきました。バブルが崩壊してから、およそ平成2年（1990年）から、本当に不景気になってきて、旅館と食堂を両立させながらリピーターのお客様をずっと大事にしていました。

今度の改修のおかげで、本当にたくさんの人たちに知ってもらうことができました。メディアでも紹介してもらったりしました。これほど大々的に日本全国から皆さんにきてもらうなんて、今までなかったですからね。本当に良かったと思います。皆さんはバスで湯沢の方からきて、バス停で降りてここへ歩いて来られます。そして道を聞くわけですよ。「もう少し奥ですよ！2キロもあるんですよ！」と言うと、

I was born in Tokamachi. I moved here after getting married. It was only my third day here when the hotel received 50 guests. It was in October, the best time for maple viewing. People had walked all the way from the other end of the gorge along the hiking trail. They would say, "any room would be fine. Please let me stay overnight." I was surprised by the number of people and thought to myself, "Is this how it's going to be every day?"

Over time, fewer people came. Then in Showa 63 (1988) the tragic accident occurred. Consequently, using the trail or entering the gorge was banned without exception. Previously, a disaster in the year Showa 59 (1984) destroyed our business. We persevered through, and rebuilt the hotel in Showa 61 (1986). My husband had always wanted to run a canteen, so we later included a canteen in our hotel.

On a regular day's work, guests would begin to check out at 10am. We then prepared food at the canteen which opened at 11am. For a long time, this was the routine of our lives until the economic bubble burst. In the second year into the Heisei era (1990), the economy stagnated. We managed to continue to run our hotel and appreciated every single visit by guests.

Thanks to the Tunnel of Light, Kiyotsu Gorge has received much attention from the media. It is the first time that people from all over Japan are learning about

びっくりしますけど、でも景色を見ながら行っても
らえれば楽しいですよ。

ここのお蕎麦はとても美味しいですね。昔からのや
り方で、つなぎに山ごぼうの葉っぱを使ったお蕎麦
を出していました。結構、評判がいいですよ。ここ
の夏は涼しくて、紅葉も綺麗ですが、冬に雪の中でか
んじきをはいて歩いてみてもらったり、いろいろな
動物の足跡を見てもらうと、とても楽しいですよ！

80

上村喜子 渓谷食堂オーナー

私はここで生まれました。昭和54年（1979年）に
オープンしてから、ずっと今まで食堂をやってます。
地元のお蕎麦は生そば使っていて、山の清水で洗っ
ているから、とても美味しいですよ。ぜひ皆さん食
べにきていただきたいと思いますね。

今回トンネルを改修していただいたので、お客様が
いっぱい増えました、特に遠いところから来たお客
様、外国人の方にも来ていただいて、本当にありがた
いです。外国のお客様とコミュニケーションとまで
はいかないかもしれないですけど、でも本当に私の
作ったものをちゃんと食べてくれている様子を見て、
「あ、食べてくれた！ ありがたいわ」と思っています。

一番奥の「ライトケーブ」のところが好きです。想像
していたよりずっと綺麗で、やっぱり感動しました。
水盤に自然が映り、幻想的な感じで、とても素敵だと
思いました。角度によって撮り方もいろいろありま
すし、面白いと思っています。皆さんがインスタな
どで投稿してくれるので、私も知らない場所を見て、

this area. Visitors coming from Yuzawa would get off the
bus at the parking lot and walk here. When they asked
for directions, I told them to go farther into the gorge for
another two kilometers, which surprised many. But after
seeing the scenery, they all felt the journey was worth it.

The regional buckwheat makes delicious noodles that
uses yamagobo leaves as the binder. It is one of the most
traditional ways of making soba noodles. In summer,
cool breezes brush by. People admire the maples in
autumn. In winter, one walks in the snow with warm
boots and comes across trails of hoofprints.

Yoshiko Kamimura, Owner of the Keikoku
Restaurant

I was born here. I opened a restaurant in Showa 54
(1979) and have been running it ever since. The soba
noodle here is made using pure buckwheat and mountain
water. When you visit, be sure to try it.

The renovation of the tunnel has brought more guests
to my restaurant. Many came a long distance, and
sometimes from abroad too. Although I have not
had many exchanges with the customers, I try to pay
attention to their body language to see if they like the
food. It is always a joy to see them savoring the food
with delight.

My favorite part of the tunnel is the light cave. I
was deeply moved by what I saw. The water pond
elegantly reflected the scenery. The overall effect was
dream-like and beyond imagination. Sometimes I
would see photographs on Instagram. When shot from
different angles, the scenery also resonates differently.

「あー、こんないいところがあったんだわ」って、すごく面白いと思いましたね。ちゃんとそこまで自然を感じてくれているから、ありがたかったですね。

81

徳井靖　清津峡渓谷トンネル支配人

私は株式会社なかさとに勤めています。十日町市から清津峡のトンネルの業務委託を受けております。私は2019年の4月から支配人になりました。具体的な仕事の内容としましては、従業員の管理、トンネル清掃、業者の委託の管理、市役所との調整などがあります。ここには7時45分に来ます。そこから営業の準備をして、清津峡トンネルは8時半からオープンとなります。そこからは受付の業務をしたり、電話の応対をしたり、中を回ったりしていますね。休憩をはさみ、夕方、営業を終了後に清掃をするということになっています。

「Tunnel of Light」は芸術作品なので、運営中にいろいろなものを管理しなければならないので、仕事が増えたと思います。外人さんもたくさん来ていますが、私は皆さんのような英語は当然話せないので、英語で話しかけられたら、日本語で返しています。

私が一番好きなところは奥の「ライトケーブ」です。日々違いますし、一番いいと思いますね。正直に申し上げますと、そこまで反響があるとは思っていませんでした。改修前の3倍に近い人に来ていただいているので、本当にありがたいというふうに思っております。

Some visitors capture views that I never knew about. I thought to myself, "what a wonderful place I am in." It is inspiring to see how people experience nature in the tiniest details.

Yasushi Tokui, Manager of the Kiyotsu Gorge Tunnel

I work for the Nakasato Co. Ltd., a property service company with businesses in Tokamachi and the Kiyotsu Gorge area. I began working as a manager in April 2019. My duties include managing employees, cleaning the tunnel, undertaking commissions for other related services, and coordinating with the city government. I usually arrive at 7:45 to get ready. After the tunnel opens to visitors at 8:30, I work at the front desk, answer phone calls, and patrol the tunnel. There is also time for a break during the day. I clean up after the tunnel closes.

Because the Tunnel of Light is an artwork, maintenance is essential for it to continue to operate. It naturally increased my workload. Many foreigners came to visit. Like most locals, I do not speak English. Sometimes people talked to me in English, and I could only answer in Japanese.

My favorite part is the light cave. The view changes every day. The number of visitors has tripled since the completion of the renovation project. I was not expecting such a positive response. I am genuinely grateful for this.

82

左：藤村真美子 チャドカンオーナー
右：大庭ひとみ チャドカンオーナー

Left: Mamiko Fujimura,
Owner of the Chadkan Food Stall
Right: Hitomi Oba,
Owner of the Chadkan Food Stall

私の出身は大阪府です。大学は九州で、勤務が東京でした。働いている時に「大地の芸術祭」サポーターのこへび隊で、この地域に関わりをもつようになりました。東北の大震災があって「もう東京じゃなくて田舎に住みたい」と思うようになり、地域おこし協力隊で十日町に来たのです。清津峡は前から気になり、とても魅力的なところと思っていました。3年間の協力隊の活動を終えて、2019年4月からチャドカンという移動販売のバングラデシュカレーの販売を清津峡で始めました。大庭ひとみさんは私のパートナーです。彼女の出身は東京ですが、田舎にずっと憧れがあって、自然の中で暮らしてみたかったようです。地域おこし協力隊の募集を見て、十日町に来ています。

チャドカンはバングラデシュの言葉で「茶店」という意味です。私は十日町に来る前に、青年海外協力隊でバングラデシュに行っていました。バングラデシュカレーなら地元の素材の味をかなり生かせるのではないかと思って、地域おこし協力隊を終えたらここでお店を始めたいと考えました。

私は奥の「ライトケーブ」と途中のライトの感じが好きです。大庭さんは赤のところが好きで、宇宙のような感じがするそうです。アーティストじゃないと思いつけないような清津峡の魅力があります。この見せ方に、私はワクワクしました。以前より、お客様がたくさん増えました。平日でもたくさん訪れているという感じ、若い人がいっぱいやって来るようになったと思います。例えば、今日はスリランカの方がたまたま通りがかって、カレーを食べてくれま

I was born in Osaka and went to university in Kyushu. My first job was in Tokyo. I first learned about this area when I joined the volunteer group (koebi-tai) of the art Triennale. After the Tohoku earthquake, I wanted to leave Tokyo and move to the countryside; it became my wish. Later, I fell in love with the Kiyotsu Gorge area when I took part in the Local Vitalization Cooperator. After three years of working in the cooperator, I launched "Chadkan" in April of 2019. Chadkan is a curry truck. Ms. Oba (Hitomi Oba) is my partner. She was born in Tokyo but has always yearned for a life in the countryside, closer to nature. She also joined the cooperator and came to Tokamachi by chance.

"Chadkan" is a Bengali word meaning "tea house." Before I came to Tokamachi, I went to Bangladesh as part of the Japan International Cooperation Agency (JICA). With my experience there and from the cooperator, I planned to introduce Bengali curry cuisine to people in the Kiyotsu Gorge area and make the most of the natural flavor of regional ingredients.

My favorite sections of the tunnel are the light cave and the lighting within the tunnel. Oba likes the tangerine section the most. For her, it felt like another planet, an expression only available to artists in discovering the fascination of Kiyotsu Gorge. I was thrilled with the work. The number of visitors has increased since the new tunnel opened. There are many people even on weekdays, and we are seeing many more young people.

した。彼の日本語はとても上手で、「おいしいよ」と言ってくれました。

皆さんがいらしたら集落を散歩したり、農家と一緒に田植えをしたり、稲刈りをすることをお勧めします。季節の移り変わりを感じられ、お年寄りからいろいろ話を聞いてみたら面白いと思います。

Today, a visitor from Sri Lanka had a curry dish. He spoke Japanese fluently and complimented the food.

I recommend friends and families to walk around the market, take a stroll, sow seeds or harvest rice with farmers, feel the changing seasons, and listen to stories from the elderly.

83

観光会社のツアーガイド

A Tour Guide

今回の清津峡トンネルツアーは、私たちの会社の新しい企画です。新潟は秋恒例のシニアツアーのなかの一本になっていて、以前は八海山のロープウェイとかゴンドラなどに行っていました。このトンネルは海外の方が設計を手がけたということで話題になって、ツアーの中に組み込まれました。

私が一番好きなところは「ライトケーブ」です。あれは日本人のアイデアではないような感じがします。お客様はここに来るまでは、「何でこんなところまで、雨も降ってるのに」というように文句をいう方もいましたが、あちらでお写真を撮ってさしあげると、「とてもよかったわ」という感じになられます。インスタ映えのスポットとして、皆さんが写真を撮る舞台になっています。

トイレは「面白い！　やられました！」という感じですね。初めてここへ来る前に、事務所の方にお電話して「お手洗いありますか？」と聞いたら、「ありますけど、ちょっと使いにくいかもしれません。駐車場で済ませてください」と言われて、興味がわきました。行ってみると、実際「これはー！」という感じです。私は必ずお客様に「清津峡トンネルでは、「ライトケーブ」とトイレを必ず見てきてください」とお伝えしています。とても面白いと思います。最初に

Our latest program includes the newly renovated Kiyotsu Gorge Tunnel. We have always included places around Niigata in our autumn travel packages designed for the elderly. We used to bring people to activities like the Hakkaisan ropeway. Later, when we heard about the tunnel designed by an international architecture firm, and its growing popularity, we decided to add it to our itinerary.

The light cave is my favorite. Frankly, the concept differs from those of Japanese designers. On our way there, people complained about walking so long in the rain. But when they eventually arrived at the last platform and took pictures, they praised it. The light cave has gone viral. It has become the internet's favorite site for photo shoots.

Also, the Bubble restroom is just so fun. Before my first visit here, I called to check the availability of restrooms. The reception suggested using the restroom at the parking lot after arriving because the restroom in the tunnel can be a little "special." This made me curious, and upon visiting, I was not disappointed by what I saw. From then on, I would always recommend my guests to please visit the light cave and the tunnel

入った瞬間はドキドキしました。

改修のデザインによって、トンネルが一新されています。これまでの日本の渓谷トンネルは洞窟のような感じで、展覧会などが企画されることはありますが、モダンな雰囲気は全然ないです。この清津峡トンネルの空間は、照明の配置などが人を誘い込むような雰囲気があります。一言でいうと感動させられる、ワクワクする所だと思います。

restroom. The moment I entered these two spaces, I was amazed.

With the new design, the tunnel has taken on a fresh look. It used to be like a cave. There were also displays, events, and other attractions, but it lacked a modern feel. Now, light and spatial design have created an atmosphere that absorbs people. The transformation is more than remarkable. It is a spectacular experience to be here.

84

若い女性

妙高市から来ました。インスタで写真を見て、車で来ました。改修する前にこのトンネルに来たことはなくて、今日が初めてです。第3見晴所の赤い照明で小さな水玉がいっぱいあるようなところが可愛くて一番好きです。

A Young Woman

I drove over from Myoko city after seeing many impressive pictures on Instagram. It is my first time visiting. I enjoyed the third viewing platform; The bubbling, tangerine space was beautiful.

85

若いカップル1

私たちは新潟市から来ました。SNSを見て綺麗だなぁと思って、ぜひ訪れたいなと思って来ました。一番好きなのは最後の「ライトケーブ」です。神秘的な感じで一度見てみたかったので、来られてよかったです。私は建築業界に勤めていますので、このような現場で働きたいと思ったりします。難しい施工方法を実現させていて、自分はそこまで勇気が出ないところはあ

A Young Couple 1

We come from Niigata city. After seeing the beautiful pictures online, we wanted to visit. My favorite was the space at the end of the tunnel; it was tranquil. I came here specifically for the light cave.

I also work in the field of architecture, so I am drawn to this kind of project. The construction process must

りまけど、やってみたい気はあります。

ここの光はとても綺麗で、絶対来るべき場所だと思います。

have been particularly challenging. Even though I have always wanted to try my hand at such projects, I am not ready to face the challenge.

The lighting effect is so beautiful. Everybody should visit here at least once.

86

若いカップル2

新潟市の中央区から来ました。インスタ上でとても有名だったので、一回見てみたいと思って、来ました。改修する前は、このトンネルについては何も知らなかったです。一番好きなのは最後の「ライトケーブ」です。めっちゃ綺麗でした。トイレは少し恥ずかしくて使わなかったんですけど、外観がとても近未来的で、最初に見たときはトイレだと思わなくてびっくりしました。このトンネルがとても好きです。

A Young Couple 2

We come from the central district of Niigata city. This place is incredibly famous on Instagram. So, we wanted to come over. I only learned about the tunnel after the renovation. I enjoyed the section at the end with the pond. It was breathtaking. The restroom had a very futuristic exterior. I did not expect it to be a restroom. Though I would not want to use it, I still went inside just to see the effect. I really like this tunnel.

77-86. 撮影：岡本裕志

77-86. Photography by Hiroshi Okamoto

87

92

93

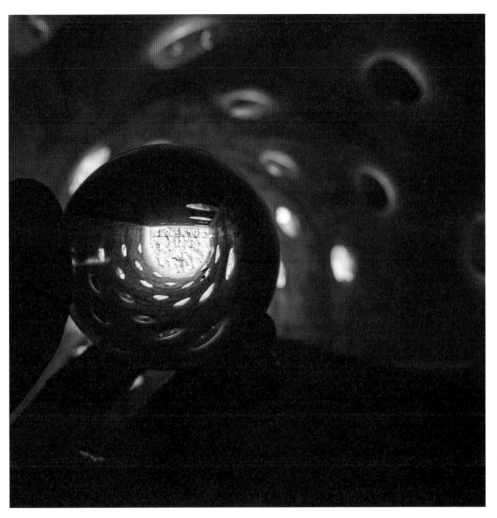

94

トンネルの光

方振寧
アーティスト、建築・美術評論家、ライター

中国のMADアーキテクツは、日本の北西部にある20年以上の歴史を持つ観光用トンネルを改修し、「Tunnel of Light」と名付けた。トンネルには、当然ながら、「入口」と「出口」がある。即ち「光のトンネル」は入口から前に進む視点からの命名であり、それを見る相手の視点、つまり出口の視点からすれば、それは「トンネルの光」だともいえるだろう。

日本三大峡谷のひとつである清津峡を見学するため、20年以上前に建造されたこの長さ約750メートルのトンネルに、いかにして再び観光客の目を引くのか？「大地の芸術祭」の主催者は、この改修プロジェクトをマ・ヤンソンと彼の事務所のMADに依頼した。MADはトンネルの中にある4か所の見晴所のうち、3か所を改修した。トンネルの第4見晴所、即ち一番奥の見晴所「ライトケーブ」は、メディアへの露出が最も高く、地元を訪れる訪問者の数は倍増した。多くの観光客を魅了した「光のトンネル」の秘密はどこにあるのだろうか？

「鏡園」から「ライトケーブ」へ

恐らくデザイナーとキュレーター以外に、マ・ヤンソンがかつて「鏡園」という作品を考案したことを知る人はいないだろう。2010年、私はマ・ヤンソンが考案した「鏡園」のプランをヴェネツィア・ビエンナーレ国際建築展の中国館に提出した。それは場に詩的な庭を与えるコンセプトだったが、残念ながら実現せず夢のような記憶のみにとどまっている。

それはどのようなプランだったのか？「ライトケーブ」とはいかなる関係なのか？

「鏡園」は、ミラー反射の原理を利用し、ミラーの園林を作ることで、ヴェネツィアの中国館に以前からあるヴァージンガーデンを更に輝かせる。自らの量

Afterword: the Light of the Tunnel

Fang Zhenning
renowned contemporary Japanese-Chinese
artist, art and architectural critic, freelance writer

The Tunnel of Light by China's MAD Architects is a renovated tourist tunnel in northwestern Japan with over 20 years of history. This tunnel linking multiple viewing platforms takes its name from a play on words: "sightseeing" in Chinese is guanguang, literally "viewing the lights." If "Tunnel of Light" implies directionality, from the tunnel mouth moving inward, then perhaps from the opposite perspective, from within the tunnel looking outward, we can discover the "light" of the tunnel.

Built over 20 years ago, this 750-meter-long tunnel is located in Kiyotsu Gorge, considered one of the Three Great Gorges of Japan. Faced with the question of how the disused tunnel might recapture the tourism spotlight, the organizer of the Echigo-Tsumari Art Triennial entrusted its renovation to MAD Architects. The approach of MAD was to redesign three of the four viewing platforms in the tunnel. The fourth, and final, viewing platform, dubbed "Light Cave," received the most attention from the media, helping to more than quadruple local tourism numbers. So what is the secret of this "Tunnel of Light?" How does it attract so many people to come "view the lights?"

From "Reflecting Garden" to "Light Cave"

Apart from some designers and curators, it's possible that readers do not know Ma Yansong had earlier conceived of a piece titled "Reflecting Garden." In 2010 I recommended his proposal to the China Pavilion at the Venice Biennial. Ma's concept would have imbued the site with the unforced elegance of a classical Chinese garden, but, sadly, it never came to fruition, and survives only in dream-like remembrance.

So what was this proposal? And what was behind the name, "Light Cave?"

感を物理的な視覚において消失させ、鏡面を使って
周囲の建物や風景を吸収、反射している。光の移り
変わりと共に、それは時空の鏡になっていく。マ・
ヤンソンとMADにとって、ミラーを使用して「消失」
を実現するのは、今回が初めてではない。

まだ記憶している人もいるかもしれないが、マ・ヤ
ンソンは何年も前に「フートン・バブル」で「鍼灸都
市」というコンセプトを提唱している。針で経穴に刺
激を与える鍼灸治療のように、慎重な介入の姿勢と
わずかな操作で都市と場の更新を実現するのがコン
セプトの本質である。これらの「バブル」にも鏡面反
射素材を使用し、それにより、既存の環境との不調和
は全くなく、むしろ場所を生き生きとさせたといえ
るだろう。

建築か？　アートか？

大地の芸術祭には、アート作品を実現するために必
要な建設プロジェクトも含まれ、さらにいえば建築
物そのものをもアートを伝える手段としている。私
たちが目にする作品は最終的にはアートであり、
「Tunnel of Light」は素晴らしいランドスケープアート
ワークだと、私は思っている。この作品の最大のポイ
ントは、限られた内部の池の水の反射を巧みに利
用し、鏡を丸くしてトンネルの出口を外の景色を取
り入れる瞳に変え、トンネルの終点を夢の入り口と
したことだ。アートの手法は、このように現実と夢
境を転換し、天地を反転させ、時間を逆流させていく。

ユニークなアートのコンセプトを水でつくる作者
は、水の柔らかさと自由という二重の特徴を熟知し
ていることだろう。流動性とコントロール不能であ
ることに水の神秘性があり、それは、マ・ヤンソンと
MADとの建築設計で追求される有機曲線と流体曲線
と本質的に同じである。「行雲流水」という言葉を私
たちはよく使うが、即ち行く雲のような形はまた、流
れる水のような形である。「ライトケーブ」は、建築
における実物の創造を水の要素に置き換えているの
みである。それは、大胆なアイデアではあるが、将来
的には水の宮殿を建てることができるだろうか？　そ
れが実現できなければ、すなわちユートピアであり、

Ma's design would have utilized the principle of reflection
to transform Venice's Garden of the Virgin, the site of
the China pavilion, into a radiant garden of mirrors.
Absorbing and reflecting the surrounding buildings and
landscape, the mirrored surfaces of the garden would have
diminished and dissipated the volume of the site. Each
time light bends, space-time is refracted. Had "Garden of
Mirrors" been realized, it still would not have been the first
time Ma Yansong and MAD Architects employed mirrors
as a means of concealment.

Perhaps some will remember "Hutong Bubble," an urban
acupuncture concept proposed by Ma Yansong a number
of years back. The substance of it was pure acupuncture:
carefully selected points where minimal intervention
can manifest urban renewal. Despite their futuristic
appearance, the amorphous reflective bubbles did not clash
with the environment of Beijing's historic lanes, instead
bringing an infusion of vitality.

Art or Architecture?

The Echigo-Tsumari Art Triennial has always entailed
certain building projects needed to house works of art,
even going so far as to envision architecture as a vessel for
art. But, at the end of the day, buildings are also works
of art in their own right. It is in this sense that I feel that
the Tunnel of Light is a masterpiece of landscape art. The
coup de maître of this work is the exquisite placement of
a small reflecting pool to complete the semi-circle of the
tunnel's mouth, creating a pupil that absorbs the image
of the outside landscape, transforming the opening into
a portal of dream-like fantasy. This is the methodology of
art, transforming the landscape of our dreams as much as it
transforms the landscape of reality, inverting the positions
of sky and earth, reversing the flow of time.

Any artist who utilizes water to implement their vision will
be familiar with water's dual nature. Mild yet untamed,
the mystery of water lies in a character that can be at
once yielding and uncontrollable. The organic curvilinear
forms of Ma Yansong's architectural designs are much the

実現できたとすれば、それは形而学上の建築である。

幸運か？　必然か？

「ライトケーブ」は完成してすぐ、多くの称賛を得て、日本のメディアが越後妻有を報道するたび、必ず触れる作品となった。なぜこの作品がそれほど人気なのか？　これまで多くの中国人アーティストがこの大地の芸術祭に招待されたが、どれもそれほど大きな反響はなかった。マ・ヤンソンは幸運なのか？　答えはNOである。

マ・ヤンソンの初期のデビュー作「フィッシュ・タンク」は、アート的な手法で建築空間を思考していた。「インク・アイス」は建築とは関係がなく、純粋なアート作品である。建築の使命が「無から有へ」の道を歩むことであるとしたら、一方アートは「有から無へ」のプロセスに専念することであり、2本の道を並行して走る車のようではなく、まったく正反対である。もちろん、マ・ヤンソンは、何度も建築を設計しながらアートに巻き込まれてきた。アートと建築設計の実践は、彼の左右の腕のようなものである。彼はアートに対して不誠実な建築家でもなければ、アートを建築デザインの仕掛けと見なす建築家でもない。彼とアーティストの違いは、彼がアートワークにより多くの空間要素を組み込む方法、アートと風景及び自然との和合について思考していることだ。「ライトケーブ」は、このような長期にわたる思考と実践の延長線上にあり、その出現は必然である。

地球上の惑星旅行

数百メートルのトンネル改修プロジェクトでは、複数の見晴所があるが、他の場所はどのように設計されたのだろう？

一般的に、数百メートルの暗いトンネルを歩いていると、外の光が見えるだけでも驚き、喜びを覚える。その他の見晴所にいかに異なる創意を持たせるかが、新しいチャレンジである。

第3見晴所は、新星が集まる場所に訪れるようなもの

same. We might say, to use a Chinese expression, they are "like passing clouds and flowing water." We might also be reminded that passing clouds are nothing other than flowing water. The "Light Cave" operates by taking an element that emphasizing the substance of architecture, and exchanging it for water. If that is possible, could we not boldly conceive of a future architecture whose entire substance is water? A water palace? If that lies beyond our ability, then let it be a Utopia. But if it is possible, it would be metaphysical architecture.

Luck or Necessity?

Praise for the Light Cave began pouring in from the moment it was completed. Print media dubbed it the "must-see" attraction of the Echigo-Tsumari festival. But why did this work of art elicit such a favorable reaction? Many Chinese artists have exhibited their works at the festival, but none have been met with such enthusiastic acclaim. Is Ma Yansong simply the lucky one? The answer is unequivocally, "No."

Ma Yansong's design debut, "Fish Tank," employed the methodology of visual art to examine architectural problems concerning space. Strictly speaking, "Ink Ice," another of his works, was pure art, with no connection to architecture. The mission of architecture is the progression from nothing to something, from emptiness to substance, while the visual arts are dedicated to the reverse process. They are two cars on the same road, racing apart in opposite directions. Of course, Ma Yansong often carries on a secret love affair with art within his architectural designs. Artistic creation and architectural design are his left and right hands. He's not one of those architects who issues empty pronouncements about art, nor does he use art as a gimmick to dress up mere buildings as architecture. But he differs from other artists by thinking more about how to bring spatial elements into his art, and how to wed art to the natural landscape. Merely an extension of this long-running train of thought and creation, the Light Cave was not luck. It was all but inevitable.

Interstellar Voyage on Earth

で、半円形のトンネルの壁には、内外部の光景を相互に映し出すランダムな円形の鏡が配置されている。これらのトンネルの鏡は穴のように感じられ、無重力空間のなかの遊星のように見え、またSF映画の暗黒の銀河をゆく宇宙艦隊のようにも見える。

自然に囲まれた国立自然保護区の日常空間が、暗い夢のような宇宙体験の空間に変容するのはいささか唐突だろうか？　そう。しかし、それは間違いなくディズニーランドのような娯楽ではなく、万国博覧会における新事物との出会いでもない。これは田舎から遠く離れた都市の人々を山、崖、滝の前に引き付ける計画の、いわばフリープランツアーである。実際、それだけではない。トンネルに入った後、旅の価値があると感じられるのは、地球上で惑星間旅行を体験できるからである。

マ・ヤンソンと彼のチームが制作した作品は、なぜ常に優れ、人を驚かせるのだろう？　これは現代中国建築の分野における建築とアート間の論争ではなく、想像力の勝負である。有名な英国の起業家にある名言がある。「地球上の最後の資源は想像力である」

これはマ・ヤンソンの出現とMADの発展に関係している。信じるかどうかはあなた次第だ。

The Light Cave wasn't the only viewing platform to be renovated in the 750 meter long tunnel. So what about MAD's designs for the others? After walking through hundreds of meters of darkness, any glimpse of light from outside can be an extraordinary surprise. Thus, the next challenge of the project was to differentiate the platforms, so each presented viewers with a novel experience.

The third viewing platform is a gathering of newborn stars. Irregular circles open in the walls of the semi-circular tunnel, reflecting the outside environment, each other, and the interior space. The mirrors can seem to be holes, or stars floating in the weightless expanse of space, or a fleet of starships flying in formation through the darkness of the Milky Way.

If visitors were asked leave a nature preserve, and immediately enter the dream-like darkness of interstellar space, the change of scene would perhaps have been too sudden. This isn't Disneyland, after all. Nor is it some experiential encounter at the World Expo. One could say it's a meticulously arranged exhibit that transports nature-deprived city-dwellers to the foot of a waterfall, to the edge of a soaring cliff, to the depths of the mountains—but actually, that's not it either. The undeniable value of a visit to this viewing platform is you will experience an interstellar journey even as you remain rooted in the beauty of this green Earth.

Why are the works of Ma Yansong and MAD Architects so successful, so unexpected, and always so far ahead of the pack? They hold their leading position in contemporary Chinese architecture not by competing in the realms of art and architecture, but by succeeding or failing on the merits of imagination. A famous British entrepreneur once said that imagination is the last untapped natural resource. If that is true, then I say the rise of Ma Yansong and the ongoing success of MAD Architects is inextricably linked to this fact. Whether you believe it or not is up to you.

付録　　　　　　　　　Appendix

感覚こそ真実
オラファー・エリアソン & 馬岩松

2010年4月4日–2010年6月20日
ユーレンス現代美術センター（UCCA）
中国、北京
キュレーター：ジェローム・サンス、郭暁彦
作品タイプ：インスタレーション

Feelings Are Facts
Olafur Eliasson & Ma Yansong

Ullens Center for Contemporary Art (UCCA)
Curators: Jérôme Sans, Guo Xiaoyan
Beijing, China
Type: Space Installation
2010.04.04-2010.06.20

オラファー・エリアソンと馬岩松は、私たちが通常用いている空間認知方法に挑む。視覚は方向ナビゲーションにおいて主要な役割を果たすが、このインスタレーションが、視覚による認識を制限し、不安の感覚を引き起こすことにより、オーディエンスは他の知覚によって環境を認識することになる。

Olafur Eliasson and Ma Yansong challenge to the spatial definition of our daily life. Vision functions works as our primary perception for navigation, but this expansive installation induces insecurity on visitors by radically reducing its visibility, thereby suggesting the need to invent new models for perception.

インク・アイス

2006年
中国、北京
作品タイプ：インスタレーション
素材：水、松煙墨
サイズ：9 x 9 x 9（feet）

Ink Ice

2006
Beijing, China
Type: Installation
Material: Water, Chinese ink
Size: 9 x 9 x 9 (feet)

インク・アイス（墨氷）は一辺9フィート、重さ27トンの黒い氷の立方体で、氷はさまざまな濃度の墨から成る。作品は開幕式当日の早朝、中華世紀壇の広場に設置され、3日間日光と風に晒されて氷は溶けていった。

3日間で、地面には自然に流動した黒色の跡が残り、物体は姿を消し、その抽象的な記号も消失し、ただ時間の痕跡と墨の跡に、無限の想像の余地が残された。

In the summer of 2006, MAD placed a 9 × 9 × 9 feet cube of ink-ice in the plaza of the China Millennium Monument, and left it to melt under the sun and wind for three days. As the temperature warmed up, the solid entity had entirely disappeared. The only residue left behind was the charcoal black imprint on the ground. The abstract symbols vanished, and a space for imagination is all that remains.

フィッシュ・タンク

2004年
アメリカ、ニューヨーク
ニューヨーク建築連盟よりヤングアーキ
テクツ賞受賞（2006年）
サ　イ　ズ：300(L) x 300(W) x 400(H)
(mm)

Fish Tank

2004
New York, USA
2006 Architectural League of New York's
annual Young Architects Award
Size: 300(L) x 300(W) x 400(H) (mm)

我々は探している／都市での魚たちの空間を
人と魚の関係を／置き換えなければならない
主体は魚となり／空間は分裂し始める
四角い箱は形を失い／質を伴わない空間は崩壊して
ゆく
機械の時代は終わった
四角い箱に挑む／主流の文化に敵対するのではない
文化の真の所有者は一般の人々である
彼らにより視線を注ぐべきである／より高い自主性を

We are seeking / the living space for fish in the city,
And the conditions of human beings and the fish / have
to be inverted:
The fish should dominate. / The space begins to split,
Cubic boxes have melted down. / The collapse of low
quality cubic space
Marks the end of the machine era.
Contradiction with square boxes/ does not signify we
are against the mainstream culture,
The civilian own the ideal of mainstream culture,
They deserve more attentions/ /and independence.

フートン・バブル218号

2015-2019年
中国、北京
作品タイプ：四合院リノベーション

フートン・バブル32号

2008-2009年
中国、北京
作品タイプ：四合院リノベーション

Hutong Bubble 218

2015-2019
Beijing, China
Type: Courtyard Renovation

Hutong Bubble 32

2008-2009
Beijing, China
Type: Courtyard Renovation

2006年、MADはヴェネツィア建築ビエンナーレに未来の北京の構想「北京2050」を出展した。そのなかの「フートン・バブル」において、旧市街のリノベーションにあたり、必ずしも取り壊して再建する必要はなく、時空を超えて存在するかのような「バブル」を新たに追加することで、磁石のように地域コミュニティの生活を更新し、コミュニケーションを活性化できると提案した。

2009年、MADは最初の「フートン・バブル」を北兵馬司胡同32号において実現した。これは、トイレと屋上テラスに続く階段を増築したもので、まるで宇宙からやって来た小さな生命体のように見え、歴史、自然、未来がひとつの幻想の世界のなかに存在しているようである。

2019年、「フートン・バブル218号」が竣工した。MADは北京の前門東区にある清朝末期の四合院の修復、リノベーションを行った。3つの異なるかたちの「バブル」を追加した。MADはこのプロジェクトで旧市街再活性化計画に対し、変わらないこと、スケールの細密化、都市への刺激、エリアの精神の保存の4原則を打ち出した。アートが地域コミュニティーに活性をもたらし、旧市街に、新旧、伝統と未来の共存する新たな対話空間を創り出した。

In 2006, at the Venice Architecture Biennale, MAD proposed ideas "Beijing 2050" for the future ambitions of Beijing. MAD conceived a network of metallic bubbles to be cultivated in Beijing's historic neighborhoods. Inserted into the city's existing urban fabric, they were envisioned to attract new people, activities and resources back to these aging and neglected communities.

In 2009, MAD's first hutong bubble was realized in Beibingmasi hutong. Hutong Bubble 32 provides a toilet and a staircase that extends onto a roof terrace for a newly renovated courtyard house. Its shiny exterior renders it an alien creature, and yet at the same time, it morphs into the surrounding wood, brick, and greenery. The past and the future can thus coexist in a finite, yet dream-like world.

In 2019, MAD has completed the project "Hutong Bubble 218," a restoration and reconstruction of a traditional courtyard house that dates back to the Qing Dynasty. It is a continuation of MAD's exploration into urban renovations as a means of revitalizing the Chinese capital's old neighborhoods that have been confronted with degradation and demolition as a result of rapid development. MAD focused on four principles: historic preservation, urban regeneration, re-energizing the community spirit and small-scale intervention. MAD's renovation demonstrates how small-scale, artistic interventions can provide new spaces and programs for these adverse areas, creating a dialogue between the old and the new.

浮遊する島——ニューヨーク ワールドトレードセンター再建計画　　　　　*Floating Island – Rebuilt WTC*

2001年　　　　　　　　　　　　　　　　　　　　　　　　　　2001
アメリカ、ニューヨーク　　　　　　　　　　　　　　　　　　　New York, USA
作品タイプ：都市構想　　　　　　　　　　　　　　　　　　　　Type: Urban Concept

2001年、911同時多発テロ事件はアメリカに多大な傷を与え、ワールドトレードセンターの再建計画については当時さまざまな業界で幅広く議論された。私たちは、ミュージアムや記念碑を建てるのではなく、全く新たなデザインの建築物を設計することで、モダニズムが提唱する「機械の美学」や「垂直都市」などの伝統的なアプローチを脱却すべきだと考えた。つまり、すべての出発点はひとつの概念に起因する——過去を乗り越え、未来のこの街の発展を展望すること。

新たなワールドトレードセンターは、無機質な機械に占有された空間ではなく、生命を感じられる複合体とした。浮遊する島はワールドトレードセンター跡地上空を覆う景観である。マンハッタンのスカイラインの上を漂う雲のように、金融街の閉鎖的な孤立状態を変え、ミッドタウンとウォーターフロントを結び付け、金融街の心臓部と融合させ、都市生活に活力を引き込む。

In 2001, the September 11 attacks left an unforgettable shock on New York City, and plans were made for the former site of the World Trade Center to be rebuilt as a memorial of the historic event. Envisioned by Ma Yansong, Rebuilt WTC is a programmed landscape flowing above the WTC site: a horizontal urbanscape joining the surrounding buildings and reclaiming public arena in the heart of the city. This new organizational structure diminishes the machine aesthetic and social divisions of the modern era. The greatest monument we can offer is a renewed public space and architecture of cohesion.

MADについて

MADアーキテクツは中国の建築家マ・ヤンソンにより2004年設立され、マ・ヤンソン、ダン・チュン、早野洋介により主宰される建築事務所である。東洋の思想を取り入れ建築の未来を模索し、人と自然、環境との情緒的関係を創造し、建築文化の在り方を探求している。

MADのプロジェクトは、都市計画、都市総合体、公共建築、博物館、オペラハウス、コンサートホール、住宅、都市リノベーションからアート作品にまで及び、中国、カナダ、イタリア、フランス、オランダ、日本、アメリカ等、世界各国にて展開している。2006年MADはカナダのアブソリュート・タワーの国際コンペで一等を獲得、ミソサガ市に「モンロー・タワー」を設計し、海外の重要建築プロジェクトを勝ち取った中国初の建築事務所となった。2014年MADはルーカス・ミュージアム・オブ・ナラティブ・アートの国際コンペに勝ち、海外の重要な文化プロジェクトを勝ち取った中国初の建築事務所となった。MADの文化プロジェクトは、オルドスミュージアム (2011年竣工)、ハルビンオペラハウス (2015年竣工)、Tunnel of Light (2018年竣工)、中国フィルハーモニー・コンサートホール (建設中)、義烏オペラハウス、ロッテルダムFENIX移民博物館 (建設中)、海口雲洞図書館 (2021年竣工)、深セン湾文化広場 (建設中) 等。その他都市プロジェクトは、日本のクローバー・ハウス (2015年竣工)、朝陽公園プラザ (2017年竣工)、四合院幼稚園 (2020年竣工)、ヤブリ起業家コングレスセンター (2021年)、嘉興駅 (2021年竣工)、衢州スポーツキャンパス (建設中)、南京ヒマラヤセンター (建設中) 等。

建築の実践と同時に、MADは書籍出版、建築展覧会、学術講座やレクチャーを通して、建築、文化、アートに対する思考を記録し探求している。MADの出版物は、"Mad Dinner","Bright City","MA YANSONG: From (Global) Modernity to (Local) Tradition","Shanshui City","MAD X","MAD Rhapsody","Tunnel of light"等。MADは国内外の文化芸術機構が主催する重要な展覧会に参加している。2019年フランスのポンピドゥーセンターにて、収蔵するMADの建築模型パーマネントコレクションの個展"MAD X"を開催。2014年UCCAにて「山水都市」個展開催。2010年UCCAにてアーティストのオラファー・エリアソンと共同で「感覚こそ真実」展開催。2007年デンマーク建築センターで個展"MAD in China"開催。MADは「ヴェネツィア建築ビエンナーレ」及び「ミラノデザインウィーク」に複数回出展。ヴィクトリア&アルバート美術館 (ロンドン)、ルイジアナ近代美術館 (コペンハーゲン)、MAXXIイタリア国立21世紀美術館 (ローマ) 等での展覧会に出展。パリのポンピドゥーセンターや香港のM＋ミュージアムは、MADの数々の建築模型をパーマネントコレクションとして収蔵している。

MADアーキテクツは、北京、ロサンゼルス、ローマ、嘉興に事務所を構える。

About MAD Architects

Founded by Ma Yansong in 2004, MAD Architects is led by Ma Yansong, Dang Qun, and Yosuke Hayano. It is committed to developing futuristic, organic, technologically advanced designs that embody a contemporary interpretation of the Eastern affinity for nature. With a vision for the city of the future based in the spiritual and emotional needs of residents, MAD endeavors to create a balance between humanity, the city, and the environment.

MAD's projects encompass urban planning, urban complexes, municipal buildings, museums, theaters, concert halls, and housing, as well as art and design. Their projects are located in China, Canada, France, Italy, Japan, the Netherlands, and the United States. In 2006, MAD won the design competition for the Absolute Towers in Mississauga, Canada. Through this, MAD became the first Chinese architecture firm to build a significant high-rise project abroad. In 2014, MAD was selected as principal designer for the Lucas Museum of Narrative Art in Los Angeles, USA, becoming the first China-based architecture firm to design an overseas cultural landmark. MAD's signature cultural projects include Ordos Museum (2011, China), Harbin Opera House (2015, China), Tunnel of Light (2018, Japan), China Philharmonic Concert Hall (under construction), Yiwu Grand Theater (under construction), FENIX Museum of Migration in Rotterdam (under construction), Cloudscape of Haikou (2021, China), and Shenzhen Bay Culture Square (under construction). Other urban projects include the Clover House kindergarten (2015, Japan), Chaoyang Park Plaza (2017, China), Yabuli Entrepreneurs' Congress Center (2021, China), Jiaxing Train Station (2021, China), Quzhou Sports Campus (under construction), and Nanjing Zendai Himalayas Center (under construction), among others.

While practicing architecture, MAD documents and discusses its reflections on architecture, culture, and arts through publications, architectural exhibitions, as well as academic lectures and presentations. MAD's publications include *Mad Dinner*, *Bright City*, *MA YANSONG: From (Global) Modernity to (Local) Tradition*, *Shanshui City*, and *MAD X*. MAD has organized and participated in several contemporary art and design exhibitions, including *MAD X*, a solo exhibition at the Centre Pompidou in 2019; *Shanshui City*, at UCCA in 2014; *Feelings are Facts*, a spatial experience exhibition with artist Ólafur Eliasson at UCCA in 2010; and MAD in China, a solo exhibition at the Danish Architectural Center, Copenhagen in 2007. MAD has participated in significant exhibitions at several iterations of the Venice Architecture Biennale and Milan Design Week. MAD has also participated in exhibitions at the Victoria and Albert Museum (London), the Louisiana Museum of Modern Art (Copenhagen), and MAXXI (Rome). An array of MAD's architecture models have been acquired by the Centre Pompidou and M+ Museum (Hong Kong) as part of their permanent collections.

MAD has offices in Beijing (China), Jiaxing (China), Los Angeles (USA), and Rome (Italy).

マ・ヤンソン
MADアーキテクツ創始者、パートナー

マ・ヤンソンは中国北京出身で、新しい世代の建築家を代表する一人として賞賛され、海外の重要建築プロジェクトを初めて勝ち取った中国の建築家である。建築の未来を探求し、都市の密度、機能と山水の境地を融合させ、人と自然の情緒的つながりを新たに確立することで、全く新しい、人の精神が中心に据えられた都市文明時代を目指している。2002年に設計した「浮かぶ島」に始まり、「モンロー・タワー」、ハルビンオペラハウス、フートン・バブル、朝陽公園プラザ、中国フィルハーモニー・コンサートホール、衢州スポーツキャンパス、義烏オペラハウス等の想像力に溢れた作品に至るまで、世界中で未来の理想的な空間の在り方を提案し実践している。2014年、マ・ヤンソンはアメリカのルーカス・ナラティブアート・ミュージアムの首席建築家となり、海外の重要な文化的建築プロジェクトを勝ち取った中国初の建築家となった。同時に、国内外の個展、出版物、アート作品を通して、都市と建築の文化的価値を探求している。

2006年ニューヨークアーキテクチュア・リーグよりヤング・アーキテクツ賞受賞。2008年イギリスのICON Magazineで「最も影響力を持つ若手建築家20人」に選出。アメリカのFast Company Magazineで「建築界で最もクリエイティブな10人2009」及び「ビジネス界で最もクリエイティブな100人2014」に選出。2010年「RIBAインターナショナル・フェローシップ」を授与された。2014年世界経済フォーラムで「ヤング・グローバル・リーダー（YGL）」に選出。

マ・ヤンソンは北京建築工程学院（現北京建築大学）建築学学士、アメリカイエール大学建築学修士を取得。清華大学、北京建築大学、アメリカ南カリフォルニア大学客員教授。

Ma Yansong, Founder & Principal Partner

Beijing-born architect Ma Yansong is recognized as an important voice in the new generation of architects. He is the first Chinese architect to win an overseas landmark-building project. As the founder of MAD Architects, Ma leads design across various scales, with the vision to create a new balance among society, the city, and the environment through architecture. Since designing the "Floating Island" in 2002, Ma has been exploring the ideal future of living through international practices. At MAD, Ma has created a series of imaginative works, including Absolute Towers, Harbin Opera House, Hutong Bubbles, Chaoyang Park Plaza, Quzhou Sports Park, and Yiwu Grand Theater. In 2014, Ma was selected as the principal designer for the Lucas Museum of Narrative Art, which made him the first Chinese architect to design an overseas culture landmark. Parallel to his design practice, he also explores the cultural values of cities and architecture through domestic and international solo exhibitions, publications, and art works.

In 2006, Ma was awarded the "Young Architects Award" by the Architectural League of New York. In 2008 he was selected as one of the "20 Most Influential Young Architects" by ICON magazine. Fast Company named him one of the "10 Most Creative People in Architecture in 2009" and one of the "100 Most Creative People in Business in 2014." In 2010 he received the "RIBA International Fellowship". In 2014 he was awarded "Young Global Leaders (YGL)" by the World Economic Forum.

Ma holds a Bachelor's degree from the Beijing Institute of Civil Engineering and Architecture, and holds a Master's Degree in Architecture from Yale University. He has been an adjunct professor at Beijing University of Civil Engineering and Architecture, Tsinghua University, and the University of Southern California.

ダン・チュン
MADアーキテクツ パートナー

ダン・チュンは中国上海出身で、MADの運営管理の中核を担い、世界各国から集まる100名以上のスタッフを率いる。MADの全てのプロジェクトの実行、MADの掲げる理論と文化の推進、グローバル戦略とマネジメントの責任者である。

MADのプロジェクトの揺るぎない推進者であり執行者として、クオリティ・コントロールやチームの人員配置から効率性管理まで、全ての管理監督を行う。同時に、全てのプロジェクトの始まりから竣工までの全工程において、クライアントや協力会社等とのコミュニケーションと調整に奔走し、関係各方面の協力があって初めて成し得る、デザインと思想を最大限尊重した高度な建築の実現に尽力している。また、最先端の建築技術の動向を理解し、掌握し活用することで、MADの設計デザインを最高のクオリティで実現させている。

アイオワ州立大学で建築学修士号を取得。プラット・インスティテュート客員教授、アイオワ州立大学助教授。

Dang Qun, Principal Partner

Born in Shanghai, China, Qun Dang is the core leader of MAD, managing a firm of over 100 architects from around the world. She is responsible for all of MAD's architectural projects, theoretical and cultural development in the firm's practices, and global strategic management and operations.

Dang is involved at every level of MAD's creativity and operations. She oversees all project execution, including the planning and deployment of teams. Her extensive experience in design and construction allows her to actively contribute at all project stages, from inception to completion. Dang is at the forefront of client communication and collaboration from initial concept to final product, and ensures all parties are aligned in delivering a unified design vision. She also leads MAD to actively engage and investigate cutting edge building techniques and technology, facilitating MAD's ever-growing vision and standards.

Dang holds a Master's degree from Iowa State University. She has held a visiting professorship to the Pratt Institute, an assistant professorship at Iowa State University.

早野洋介
MADアーキテクツ パートナー

早野洋介は日本愛知県出身で、日本の一級建築士である。MADのパートナーとして、MADの全てのプロジェクトの監督管理と指示指導を担う。確固たる専門的経歴と、細部にわたる高度なレベルのコントロール能力により、プロジェクトごとに異なるあらゆるスケール：建築的スケールから都市スケールに至るまで、MADの設計理念を徹底して実現させるべく、チームを統率する。コンセプトスケッチ、テクニカルドローイングから最終的な建築形態まで全ての過程において一貫して、ユニークかつ敷地の条件に対応した建築的な解答を模索し、デザインコンセプトを完璧に実現し、MADの基準に合致する建築を実現させている。

2000年早稲田大学理工学部材料工学科卒、2001年早稲田大学芸術学校卒、2003年ロンドンAAスクール建築学修士を取得。2006年ニューヨークアーキテクチュア・リーグよりヤング・アーキテクツ賞、2011年デザイン・フォー・アジア賞、くまもとアートポリス推進賞選奨受賞。2008-2012年早稲田大学芸術学校非常勤講師、2010-2012年東京大学外部講師。2015-2019年ロンドンAAスクール外部有識者審査員。

Yosuke Hayano, Principal Partner

Yosuke Hayano was born in the Japanese region of Aichi and is a first-class registered architect in Japan. He oversees all design works at MAD, directing each design team to seamlessly materialize MAD's philosophy, from concept sketch to technical drawing to final architectural form. Yosuke also oversees MAD's design language across all scales, from the human scale to the architectural and urban scale, thus implementing MAD's unique, site-specific architectural response for each project.

Yosuke received his Bachelor's degree in Materials Engineering from Waseda University in Tokyo in 2000. He gained his Associate degree in Architecture from the Waseda Art and Architecture School in 2001, and his Master's degree in Architecture from the Architectural Association of London in 2003. Yosuke has been the winner of several high-profile awards in his career to date, including the Architecture League of New York Young Architects Award (2006), the Design for Asia Award (2011), and the Kumamoto Artpolis Award (2011). He served as a visiting lecturer at the Waseda Art and Architecture School from 2008 to 2012, and at Tokyo University from 2010 to 2012. In addition, he acted as an external examiner at the Architectural Association of London from 2015 to 2019.

Ma Yansong (left), Dang Qun (center), Yosuke Hayano (right). Photography by Greg Mei.

概要
場所：越後妻有　日本
時期：2018年、2021年

チーム
デザインチーム：マ・ヤンソン、早野洋介、ダン・チュン、藤野大樹、宮本一志、石神勇樹、秦健二

現地協力建築設計事務所：株式会社グリーンシグマ

MAD編集チーム：
フィオナ・チー・ズーイン、タミー・シエ、ジャン・リーミン、シャオ・イーシュエ、グー・シャオヤン

日本語訳：
早野香苗、フー・ジャーリン

日本語校正：
原口純子

英語訳：
リン・ヤン、Josh Dyer

英語校正：
Niall Patrick Walsh

グラフィックデザイン：
One Thousand Times

Overview
Location: Echigo-Tsumari, Japan
Time: 2018, 2021

Team
Design Team: Ma Yansong, Yosuke Hayano, Dang Qun, Hiroki Fujino, Kazushi Miyamoto, Yuki Ishigami, Kenji Hada

Executive Architect: Green sigma Co., Ltd.

MAD Editing Team:
Fiona Qi Ziying, Tammy Xie, Zhang Liming, Shao Yixue, Gu Xiaoyan

Japanese Translation:
Kanae Hayano, Hu Jialin

Japanese Proofreading:
Junko Haraguchi

English Translation:
Lin Yan, Josh Dyer

English Proofreading:
Niall Patrick Walsh

Graphic Design:
One Thousand Times

馬　岩松　　光のトンネル

2021年11月30日発行

編集	MADアーキテクツ
発行者	北川フラム
発行所	株式会社現代企画室
	〒150-0031
	東京都渋谷区桜丘町29-18
	ヒルサイドテラスA-8
	TEL 03-3461-5082
	FAX 03-3461-5083
	http://www.jca.apc.org/gendai/
印刷・製本	三永印刷株式会社

Ma Yansong,　Tunnel of Light

Date of Publication: November 30, 2021

Edited by	MAD Architects
Published by	Fram Kitagawa
Publisher	Gendaikikakushitsu Publishers
	A-8, Hillside Terrace
	29-18, Sarugaku-cho, Shibuya-ku
	Tokyo, 1500033 Japan
	TEL +81-3-3461-5082
	FAX +81-3-3461-5083
	http://www.jca.apc.org/gendai/
Printed by	Sanei Printery Co., Ltd.